Against the Crisis

Ståle Holgersen is a senior lecturer in human geography at Stockholm University, Sweden. He is a member of two research collectives: the ecosocialist Zetkin Collective, which produced *White Skin, Black Fuel* (Verso, 2021), and housing research collective Fundament, which wrote *Kris i Bostadsfrågan* (Daidalos, 2023). This is his first monograph in English.

Against the Crisis

Economy and Ecology in a Burning World

Ståle Holgersen

London • New York

To Rosa

This English-language edition first published by Verso 2024

First published as *Krisernas tid. Ekologi och ekonomi under kapitalismen*

1 3 5 7 9 10 8 6 4 2

Verso
UK: 6 Meard Street, London W1F 0EG
US: 388 Atlantic Avenue, Brooklyn, NY 11217
versobooks.com

Verso is the imprint of New Left Books

ISBN-13: 978-1-80429-380-5
ISBN-13: 978-1-80429-381-2 (UK EBK)
ISBN-13: 978-1-80429-382-9 (US EBK)

British Library Cataloguing in Publication Data
A catalogue record for this book is available from the British Library

Library of Congress Cataloging-in-Publication Data
A catalog record for this book is available from the Library of Congress

Typeset in Minion by Hewer Text UK Ltd, Edinburgh

Printed and bound by CPI Group (UK) Ltd, Croydon CR0 4YY

Contents

Preface

My life motto is that the poorer a person is, and the less power they have, the more respect they deserve – and vice versa. This book is based on this life motto. It is written in solidarity with all those affected by the crises of capitalism and aims to be a merciless critique of the ruling class that creates these crises.

The scope of this book is very broad, but also very narrow. The fact that it spans a wide range of economic, ecological, social and political fields – across time and space – is a deliberate choice. According to socialist biologist Richard Levins, the complexity of, say, a 'world syndrome' can feel overwhelming. But the really big scientific disasters, Levins says, have been caused precisely by people solving one problem at a time and not seeing the monster they created from the other end: pesticides have increased pest problems; antibiotics have created new resistant bacteria; economic development has created hunger.[1] For this very reason, we should not be intimidated by the fact that capitalist crises force us to think broadly. In context of the economy alone, crisis policy is about restoring growth and profits. If successful, such policies lead to more ecological crises. In the context of ecology, the most effective way to stop, for example, the rise in temperature would be to put the emergency brakes on and immediately reduce all production and

1 Richard Lewontin and Richard Levins, *Biology under the Influence: Dialectical Essays on Ecology, Agriculture, and Health* (New York: Monthly Review Press, 2007).

consumption, which – all else being equal – would inhibit economic growth and lead to depression, unemployment and massive poverty. For this reason, Andrew Sayer calls this a *diabolical crisis*: the solution to one creates the conditions for another.[2] The crises appear to exist in parallel universes. One (the ecological crisis) is a consequence of capital accumulation, while the other (the economic crisis) is a consequence of capital accumulation coming to a halt. Only with a broad approach, in which we analyse economic and ecological crises on different levels of abstraction, can we see how they are different, but they also come out of the same system. They are two different capitalist crises. In other words, economics is too important to be left solely in the hands of economists; ecology is too important to be left solely in the hands of people who 'love nature'; and class is far too important to be left solely in the hands of sociologists. The nature of the crises forces us to think broadly and dialectically.

At the same time, the focus of this book is quite narrow. Very few concrete crises are treated with the depth they deserve. Modern capitalist crises are complex phenomena, and I discuss several of them mainly from an overall and general perspective. Issues such as racism, nationalism and fascism are discussed in general formulations, and only from the vantage point of crises. There are large fields that I discuss only indirectly and fragmentarily: everything from gender, alienation and psychology to geopolitics and imperialism falls into the shadows.

This book is about ecological and economic crises. We live in a time of crises in several spheres and areas – social, political and geopolitical, social reproductive, hegemonic, ideological – so why focus on just these? There are three arguments for my delimitation. The first is simple: this is where my expertise lies, this is where I believe I have something useful to contribute. Analyses that consider gender or psychology in relation to the crises of capitalism can be extremely important and interesting – I will have to leave it to others who know those areas better to follow such trails. The second argument is that these two perspectives are absolutely crucial for understanding the path the world will take in the coming decades and for understanding the world our children and grandchildren will live in. In the midst of these crises, it seems hard to say anything about the future. But one thing seems certain. The children

2 Andrew Sayer, *Why We Can't Afford the Rich* (Bristol: Policy Press, 2015), ch. 21.

born today who live to be seventy or eighty years old will die in a very different world than the one they were born into – largely due to ecological and economic crises.

My third argument for analysing these two crises is that they reveal something about capitalism. At first glance, this book seems to analyse capitalism in order to understand crises. My ambition is to do the reverse, as well. When we understand crises as necessarily existing as both underlying processes and actual events – as I argue we should – we must explore the multifaceted tensions between the abstract and the concrete; structures and actors; economy and history; and the particular and the general. Crises can help us understand the world.

The book was written in Sweden by a Norwegian. It is set in Scandinavia but is not just about Scandinavians. The approach is global. Not because I seek to explain what is happening in every corner of the world, but because the crises themselves have become global. If I want to understand the crises that unfold outside my window in Stockholm, I cannot escape the global context.

I started writing this book in autumn 2019. The intention was to write something short and easy to read about the climate crisis and economic crises. Then came Covid-19 and a home office, and the book became qualitatively deeper and quantitatively longer. Some sections of the book have become relatively complex, but my ambition is to write simply without simplifying; to be politically relevant rather than philosophically interesting.

Writing about crisis in the midst of an ongoing crisis presented some challenges. Crises are turning points and critical situations, but also times when new trends quickly disappear. It is an intellectual challenge to follow and write about a crisis without either ignoring or overemphasising recent events. Only within the few months between the Swedish and English editions, Russia attacked Ukraine, inflation became a major concern, the so-called energy crisis and cost-of-living crisis were on everyone's lips, and Israel started a genocidal war on Gaza. God knows what the world will look like when you read this! I hope the book can help you and others to better understand – and stop – tomorrow's crises.

All books are social products, and a number of people need to be thanked. First, thanks to Roya Hakimnia for reading and commenting on the book throughout the writing process. It was an intellectual privilege to experience, discuss and even write about the corona pandemic with you. Many thanks to Andreas Malm for discussions throughout the process, and valuable feedback on the manuscript, and to Shabane Barot, Ali Esbati and Göran Therborn for reading and commenting on the manuscript in the best possible way. Thanks to Rikard Warlenius, Olof Bortz, Leandro Mulinari, Hannes Rolf, Tim Blackwell, Erik Hansson, Kate Monson, Giovanni Bettini, Athena Farrokhzad, Magnus Helgesen, Shila Ghelichkhan, Håvard Haarstad, Sylvi Endresen and Per-Anders Svärd for reading and commenting, or brainstorming on titles or cover art, or general support. Thanks to A. G. Hedvig for an inspiring Zoom seminar, and Vanja Carlsson, Carl Cassegård, Lars Henriksson, Håkan Thörn, Evelina Johansson Wilén and Carl Wilén for a great workshop at Dyrön. Thanks to numerous colleagues and comrades in Oslo, Stockholm, Malmö, Karlstad, Jönköping, Sundsvall, Tønsberg, Bergen and Uppsala for comments and discussions during and after my presentations. Thanks to friends and comrades in the Zetkin Collective and the housing research collective Fundament. Many thanks to Steven Cuzner for translation, and to Sebastian Budgen and everyone at Verso. Thanks to Nils Sjödén and Daidalos. Thanks to the Marxist Summer School (Left Party Skåne) for inviting me to talk about crises in August 2019, which made me decide to write the book. The introduction and chapter 5 were written with support from Formas (project nos. 2018-01613 and 2018-01702). The book is based on work I have done over a decade, so thanks to current colleagues and peers in Örebro, Uppsala and Stockholm, but also to friends and former colleagues in Bergen, Malmö, Lund and Oslo.

A warm and heartfelt thank-you to all my friends and family, from the west coast of Norway to the east coast of Sweden, and everyone in between. And elsewhere. You know who you are and what you mean.

Most of all, I want to thank Roya, Rosa and Kaveh. There are no limits to how much I love you. With all my heart, to infinity and beyond.

Introduction: Crises Are Not Opportunities – They Are the Enemy

When millions of people lose their jobs or houses in a matter of a few weeks, or when tens of millions have to flee the world's poorest countries every year because of climate change, most people probably conclude that *something has gone wrong*. But that is not the case. Rather, the loss of jobs, homes and lives of millions due to economic or ecological crises is because *everything has gone right.*

Crises appear as extraordinary exceptions to an otherwise functioning capitalism. But, in fact, the opposite is true. Crises are created through ordinary and everyday processes of capitalism. The processes that create prosperity and wealth also create misery and poverty – and crisis. The perpetual pursuit of profit creates gold and green forests in one place and poverty, deforestation and environmental disasters in another. If capital accumulation and the use of nature are absolutely crucial for capitalism, surely a halt in accumulation (economic crisis) and the destruction of nature (ecological crisis) must be a problem for the system? This might sound counterintuitive, but in this book we will see how crises are part of capitalism, how they come out of the system, reshape it and give it new strength. Crises are simultaneously enormous problems for many people and a necessity for the capitalist system. Without the ability to renew itself through economic crises, capitalism could never survive; without an unsustainable relationship to nature, it could not exist. The crises of capitalism are not flaws in the system or a break from any progressive modernism. They are necessary components for the system to work. Without crises, capitalism would have collapsed a long time ago.

Capitalism is the time of crises. Never before have crises been written into the DNA of the social system in such a way. Never before has an economic system depended on recurring economic crises to survive. Never before has a global metabolic rift between humans and nature led us into a new geological epoch. Nowadays, the crises come from contradictions within the system itself and they come with such force that it is at once fascinating and profoundly frightening.

The concept of crisis is already found in the writings of the ancient Greeks, but in relation to law (and thus politics), medicine and theology. The concept then fell out of fashion, and was, according to the historian Reinhart Koselleck, most likely not very central until the seventeenth century. When it was used, it was primarily in the medical sense, as the observable condition and the verdict (judicium) on the course of the disease, at the decisive point at which it becomes clear whether the patient will survive or die.[1] From the seventeenth century onwards, the use of crisis expanded to describe phenomena in politics (and war), economics, history and psychology, and, from the eighteenth century onwards, the use of the term exploded. According to Brian Milstein, its use shifted from the individual to the societal body, or *body politic*.[2] The concept took on a more radical meaning: it became an essential component of the idea of progress. Crises came to be seen as important historical moments, turning points and revolutions.

That the concept of crisis 'returned' in the seventeenth and eighteenth centuries can hardly be considered a coincidence. There were things that needed to be grasped. Turbulent times that needed to be conceptualised: bourgeois and technological revolutions, disintegration of feudal society and of the aristocracy as the ruling class, broad social movements, displacements and urbanisation, and plunderings of the rest of the world.[3] But, if the use of crisis gained ground in the turbulent advent of

1 Reinhart Koselleck, 'Crisis', *Journal of the History of Ideas* 67, no. 2 (2006), pp. 357–400.

2 Brian Milstein, 'Thinking Politically about Crisis: A Pragmatist Perspective', *European Journal of Political Theory* 14, no. 2 (2015), p. 144; Koselleck, 'Crisis', pp. 360–2.

3 During the nineteenth century, the concepts of crisis and capitalism developed very much side by side. Read, e.g., Koselleck against Raymond Williams, *Keywords: A*

modern capitalism, one could think that the concept should have disappeared once the transition to capitalism was complete. The concept was needed to explain the stormy transition to capitalism, and should no longer have been needed once the new form of production had taken hold. But anyone alive in the 2020s knows that this is certainly not the case. If anything, the concept of crisis has become increasingly important as capitalism has developed. Crises under capitalism are not decisive points at which it becomes clear whether the patient will survive or die – they are, rather, what keep the patient alive.

The fact that capitalism has crises written into its DNA has implications for the ruling class. The raison d'être of rulers in all previous eras has been – as a very basic principle – 'conservation of the old in unaltered form', where rulers retained their power by ensuring that things basically remained as they were.[4] The premise under capitalism is different. Now, the ruling class cannot survive without constantly changing itself and the world around it. This requires a continuous shaking up of all social relations, a perpetual uncertainty and constant movement. When all that is solid melts into air, those in power must embrace change incessantly in order to maintain their power: new ways of organising production and new technologies, new infrastructure and new patterns of communication, new relations with workers and nature, and new ways to legitimise their power. The ruling class must constantly change to maintain the status quo. And crises, as we will explore in this book, are excellent events for rulers who seek to both change and reproduce social relations. It might be tempting to paraphrase what Voltaire said about God: if crises did not already exist, the ruling class would have to invent them.

Vocabulary of Culture and Society (London: Fontana Press, 1983), pp. 50–2; Karl Marx and Friedrich Engels, *The Communist Manifesto* (London: Pluto Press, 2008), pp. 33–6; Karl Marx, *Capital*, vol. 1 (London: Penguin, 1976); Meghnad Desai, 'Capitalism', in Tom Bottomore, Laurence Harris, V. G. Kiernan and Ralph Miliband (eds), *A Dictionary of Marxist Thought* (Oxford: Blackwell, 1983), pp. 71–5. It is often said that Marx learned to understand capitalism by studying philosophy from Germany, politics from France, and economics from England. In Koselleck's review of the history, we can identify how the concept of 'crisis' developed precisely in these fields in these countries.

4 See Marx and Engels, *The Communist Manifesto*.

Towards a Definition: Crises as Social Paroxysms

There is rarely any correlation between the attractiveness of a concept and the precision with which it is used.[5] The concept of crisis is constantly used in the media, popular culture and research, and it has long been a key term in our political lexicon.[6] But despite – or perhaps because of – its frequent use, there is no agreed definition of crisis. Koselleck argues that the term has a 'metaphorical flexibility' and has become a catchword. In the twentieth century, there was virtually no area that was not examined and interpreted through this very concept.[7] Étienne Balibar argues that the concept of crisis has been 'trivialised to the extreme'. And Nancy Fraser argues that 'whoever speaks of "crisis" today risks being dismissed as a bloviator, given the term's banalization through endless loose talk'. In the housing sector, for example, crisis can mean anything from the saturation of the market for luxury apartments to people living in crowded, unhealthy conditions. The Austrian liberal economist Joseph Schumpeter wrote a lot about economic crises, but did not like to use the word 'crisis', while others, who do not really write about crises, use it all the time.[8]

Theorising crises has always been controversial. According to Fraser, no other genre of critical theory has been as heavily criticised as this one: '["Crisis theory"] has been widely rejected, even dismissed, as inherently mechanistic, deterministic, teleological, functionalistic – you

---

5 Jürgen Kocka, 'The Middle Classes in Europe', *Journal of Modern History* 67, no. 4 (2006), p. 783.

6 On crisis and knowledge production, see Annika Bergman-Rosamond et al., 'The Case for Interdisciplinary Crisis Studies', *Global Discourse: An Interdisciplinary Journal of Current Affairs* 12, no. 3–4 (2022), pp. 465–86; Nancy Fraser and Rahel Jaeggi, *Capitalism: A Conversation in Critical Theory* (Cambridge: Polity, 2018), p. 35; Bob Jessop, 'The Symptomatology of Crises, Reading Crises and Learning from Them: Some Critical Realist Reflections', *Journal of Critical Realism* 14, no. 3 (2015), pp. 238–71; Étienne Balibar, 'Racism and Crisis', in Étienne Balibar and Immanuel Wallerstein, *Race, Nation, Class: Ambiguous Identities* (London: Verso, 1991), pp. 217–27; Koselleck, 'Crisis', p. 399; Milstein, 'Thinking Politically', p. 141.

7 Koselleck, 'Crisis', p. 397; Milstein, 'Thinking Politically', pp. 141–2.

8 Étienne Balibar, 'From Class Struggle to Classless Struggle', in Balibar and Wallerstein, *Race, Nation, Class*, p. 156; Nancy Fraser, *The Old Is Dying and the New Cannot Be Born* (London: Verso, 2019), p. 1; Koselleck, 'Crisis', p. 397; Joseph Schumpeter, *The Theory of Economic Development* (London: Transaction, 1983).

name it.'[9] Crisis theory can range from theories that reduce all crises of capitalism to a particular tendency, to theories that drown the concept in infinite complexity. According to Bob Jessop – who rarely shies away from explaining that reality is complex – crisis is a 'polysemic word and a problematic concept and denotes multi-faceted phenomena that invite approaches from different entry-points and standpoints.'[10] Because people perceive the crises of capitalism differently, because they *are* different across time and space, and because they arise from different contradictions in capitalism, there are really almost no limits to how complex the issue can be made.

There are those who welcome ambiguity. Perhaps we should move forward without a definition, as it allows us to keep 'open what it may mean in the future'?[11] However, if we want a minimum of clarity and transparency when operating with a concept that has no agreed definition, at a time when the term is proliferating, it becomes necessary to at least make some delineations.

In various textbook definitions of the term 'crisis', it often refers to a 'very difficult situation' and/or 'turning point'. This is a good starting point. According to Koselleck's historical review, the term has its roots in the Greek κρίσις (*krisis*): to 'separate', 'choose', 'judge', 'decide', a way of 'measuring up', to 'quarrel' or to 'fight'.[12] From this semantic starting point, the term can be taken in different directions. On one hand, we cannot expect a concept like crisis to mean exactly the same thing over thousands of years. On the other, we cannot remove all historical connotations or pull new definitions out of thin air. A current understanding must be sensitive to both the conceptual history and the present-day realities. We should stick to notions of turning points and difficult situations from the textbook, and combine these with analysis of actual crises under capitalism.

From this rubric I will suggest the following understanding: modern crises are necessarily based on three elements. Crises are events that 1) come relatively quickly, 2) are embedded in underlying structures and processes, and 3) have negative effects on people or nature.

9 Fraser, in Fraser and Jaeggi, *Capitalism*, p. 10.

10 Jessop, 'The Symptomatology of Crises', p. 245.

11 See Koselleck, 'Crisis', p. 399.

12 Ibid., p. 358; see also Janet Roitman, 'Crisis', in *Political Concepts: A Critical Lexicon* 1 (2012).

These three elements are central. Without the first, we can have all sorts of continuous problems. Without the second, we can simply talk about an unfortunate coincidence, accident or chance. Without the third, we can have events that are positive, as human advancement and the like.

With modern capitalist crises, we can arguably also add a fourth element: the crises require some kind of awareness of and usually response to the situation. This criterion highlights crises as turning points, and further distinguishes crises from other phenomena such as tragedies or accidents to which we must adapt but cannot do anything about. It follows from our definition that there is always a *where* and *when* for crises. Crises can always be delimited in time and space – although there is considerable flexibility here.

Working from these three – or four – elements, it follows that we have to operate with some tensions. Crises necessarily move at different levels; between surface phenomena and underlying processes; between what is measurable and how crises are experienced; between the objective and the subjective. Crises are not just extra-semiotic events or processes that automatically create a certain outcome, and certainly not just experiences and emotions independent of underlying structures and processes.[13] In his analysis of financial crises, economic historian Charles Kindleberger argues that historians tend to see each event as unique, while economists argue that forces in society and nature behave in repetitive ways: history is particular while economics is general.[14] If we are to understand the crises of capitalism, we must engage in the difficult exercise of simultaneously analysing both what is changing within capitalism and what is more enduring, both longer processes and more spectacular events. Precisely this – understanding the connection between the particular and the general, between the general laws of movement of capital and the history of capitalism – is as difficult as it is important.

The French Marxist Daniel Bensaïd captures the relationship between underlying phenomena and surface by describing crises as antagonisms that have grown into *paroxysms* – that is, sudden attacks

13 Jessop, 'The Symptomatology of Crises', pp. 238, 245–6.

14 Charles P. Kindleberger, *Manias, Panics, and Crashes: A History of Financial Crises* (New York: John Wiley and Sons, Inc., 2000), p. 13.

or outbreaks of a known disease. Crisis denotes something that creates gaps, discontinuities and breaks in the ruling order. What distinguishes crises from religious miracles, according to Bensaïd, is that crises always come in *historical situations.* A crisis is a crisis *in* something, *out of* something, *for* something or *of* something. Jared Diamond is wrong when he argues that crises can arise from nothing.[15] Crises are never independent and autonomous phenomena that move around the world and cause this or that. Lenin argued that crises are not chance happenings or trifles, but incidents which are outward signs of a deep-rooted inner crisis.[16] In this book, crises will therefore be understood as *societal paroxysms* – that is, underlying contradictions that come to the surface as definite events.

Here, we will only examine human-made crises and primarily societal consequences. Rather than ecological crisis, perhaps I should write socio-political ecological crises. And, since economic crises are certainly never purely economic, perhaps they should be called socio-economic crises or political-economic crises.[17] For the sake of simplicity, however, we will just call the crises ecological and economic.

This book examines economic and ecological crises, but we could also apply a similar approach to other types of crises. *Political* crises are events that are rooted in and have implications for underlying political structures and processes. A political crisis might be the Social Democrats losing by landslide in an election (event), if this is related to the general disintegration of social democracy as a political project (underlying process), or when the twenty-six-year-old Tunisian street vendor Mohammed Bouazizi set himself on fire on 17 December 2010, triggering a gigantic political crisis for many Middle Eastern dictatorships. A *gender crisis*, for example, can be triggered by a shocking decision to restrict abortion rights, which would contribute to the stifling of feminist movements. Or perhaps the opposite: the social winds could actually blow in another direction and the decision on abortion rights could

15 Daniel Bensaïd, 'The Time of Crises (and Cherries)', *Historical Materialism* 24, no. 4 (2016), p. 27; Daniel Bensaïd, *An Impatient Life: A Memoir* (London: Verso, 2013), p. 81; Jared Diamond, *Upheaval: How Nations Cope with Crisis and Change* (London: Penguin, 2020), pp. 9–10.

16 V. I. Lenin, 'The "Crisis of Power"', in *Lenin Collected Works*, vol. 24 (Moscow: Progress Publishers, 1964), pp. 332–4.

17 Bensaïd, 'The Time of Crises', pp. 14, 27.

be the start of massive demonstrations that crush national conservative patriarchal structures. The murder of George Floyd triggered the Black Lives Matter demonstrations which – if the movement succeeds – could result in a *crisis of racist social structures*. I am confident that the definition of crises as social paroxysms is applicable to domains beyond the economy and ecology, but a more thorough analysis is required to understand such crises, and to identify similarities, differences, how they are related to each other and to capitalism, and more. These further domains warrant books of their own.

Events can be very different in form, size and content, and the underlying processes and structures in question can indeed also be very different. Without both events and underlying processes, we do not have a crisis. However, there may be phenomena that are similar to crises. In *The Shock Doctrine*, Naomi Klein formulates a concept that is also useful in our context. She writes that the International Monetary Fund (IMF) had a strategy for creating *pseudo-crises*. If strong-willed countries did not want to dance to their tune, the IMF could fiddle the books and create a crisis situation where action was needed.[18] The concept of *pseudo-crisis* is very appropriate even for us as a description of what on the surface appears to be a classic crisis situation but which is *not* grounded in underlying structures.

Underlying processes should also not be defined as crises unless they create relatively rapid events that come with some degree of shock. However, there are authors who argue that the shocking event is not needed at all. Danish anthropologist Henrik Vigh argues that crises should be seen as permanent phenomena. Rather than understanding crisis in context, we need to see crisis *as* context. For the world's socially marginalised and poor, crises are not temporary events, but rather something chronic and permanent. This is, of course, quite correct, given that we – like Vigh – define crisis as, for example, 'poverty and distress', or equate non-crisis with 'balance, peace and prosperity'.[19] But after such an academic innovation, we would simply need a whole new concept to talk

18 Naomi Klein, *The Shock Doctrine: The Rise of Disaster Capitalism* (London: Penguin, 2007), p. 326.

19 Henrik Vigh, 'Crisis and Chronicity: Anthropological Perspectives on Continuous Conflict and Decline', *Ethnos, Journal of Anthropology* 73, no. 1 (2008), pp. 5, 7–8. For Marx on permanent crises, see Karl Marx, *Theories of Surplus-Value*, part 2 (Moscow: Progress Publishers, 1968), ch. 17.

about crisis as an event and a decisive point, which is how most of us think about the term today. Rather than reinventing the wheel, let us stick to the starting point here. There are no permanent crises.

Searching for a Socialist Approach to Crises

What legitimises power under capitalism is called progress. Although capitalism has always produced poverty and underdevelopment, it has been legitimised by the idea that 'in time' everyone will enjoy the fruits of progress. It only takes some policy reforms, a few more rounds of investment and some more development before even poor countries become rich, and poor people in rich countries become middle class. But what happens to faith in progress when capitalism is in a sea of crises? How is capitalism legitimised when both growth and jobs are lost and companies go bankrupt? What about progress when we fear the climate of tomorrow?

The crises expose capitalism. Suddenly, we can see clearly how useless the billionaire celebrity speculator really was, how little politicians really knew about our society and how dogmatic the ruling ideology was. Crises reveal what lies behind the fine theories of market freedom and self-regulating economies. We see, in the words of Henryk Grossman, 'the chaos of the destruction of capital, the bankruptcy of firms and factories, mass unemployment, insufficient capital investment, currency crises, and the arbitrary distribution of wealth.'[20] Class interests immediately become more apparent. As quickly as ideology and pretty words about freedom and human values disappear, self-interest and pragmatism emerge. Praise for the free market falls silent when the capitalist class needs support from the state. During a crisis, we have historically seen both the flexibility of the capitalist system and the desperation of the ruling class – two phenomena that should never be underestimated.

But, if crises really expose the nature of capitalism, why haven't 200 years of recurring crises sent the system to the dustbin of history? If, as Karl Kautsky said a hundred years ago, the recurring crises are

20 Henryk Grossman, 'Marx, Classical Economics, and the Problem of Dynamics,' *International Journal of Political Economy* 36, no. 2 (Summer 2007), p. 47.

memento mori – a 'reminder of death', that is, a foretaste of capitalism's final collapse – why does capitalism appear as alive today as ever? If, as Bensaïd argues, crises threaten to blow up the whole of bourgeois society, why does capitalism seem to draw additional strength and energy from each new crisis? If, as Jared Diamond argues, crises are *moments of truth* that challenge the ideology of progress, why does the ruling class seem able to use crises precisely to advance its positions, reinforce its power, and once again create a world in its image?[21]

That intellectuals can use crises to disclose capitalism is politically cold comfort. It is an illusion that a ruling ideology must be coherent.[22] Capitalism is not driven by coherent ideologies. In fact, it is not primarily driven by ideologies at all. The crises of capitalism come with a curious double character. While crises can – in theory – help us to reveal and expose capitalism's weaknesses and problems, they are also – in the actual political economy – central to the reproduction of capitalism. Crises are a good starting point for criticising capitalism, but they also make it harder to actually overthrow the system.

The crises of capitalism come with problems even for liberals. An old liberal dream is to maintain what is considered the sunny side of capitalism – growth, progress, optimism – and to be able to control or simply get rid of permanent and recurring disasters. One can try to realise this dream in different ways, for example through active state policies and regulations (as with Keynesians and social democrats) or through privatisations and deregulations (as with neoclassical and neoliberal thinkers). Together, these schools of thought seek a world based on capital accumulation, growth and progress, where crises are controlled or eradicated.

This liberal dream has been shattered again and again throughout history. The dream of many Marxists is an inversion of the liberal dream. Here, the crises are supposed to lead to the collapse of capitalism and thus to the age of socialism. This hope has been dashed just as many times as the liberal dream of a world without crises.

21 Karl Kautsky, 'Finance-Capital and Crises', marxists.org (1911); Bensaïd, 'The Time of Crises', p. 14; Diamond, *Upheaval*, p. 7.

22 See, e.g., Stuart Hall, 'Gramsci and Us', in *The Hard Road to Renewal: Thatcherism and the Crisis of the Left* (1988), published on versobooks.com, 10 February 2017.

While Marxist and socialist theories are useful tools for understanding crises, in actually existing capitalism the system is reproduced one crisis after another. For liberals, crises are theoretical problems with political possibilities. For Marxists, crises present theoretical possibilities, but political problems.

Dangers or Opportunities?

Perhaps the most common definition of crisis comes from the thirty-fifth president of the United States. John F. Kennedy said in 1959 that the Chinese word for crisis is composed of two characters – one (危, *wei* in Mandarin) meaning 'danger' and the other (機, *ji*) meaning 'opportunity' – and this great wisdom has been repeated innumerable times. In a modern take on the climate crisis, Al Gore said in 2015, 'We all live on the same planet. We all face the same dangers and the same opportunities; we share the same responsibility for charting our course into the future.'[23] The idea that we all face roughly the same opportunities and dangers in economic crises is simply wrong. In the case of climate change, the same statement becomes morbid. (According to Victor H. Mair, a professor of Chinese language and literature, Kennedy was wrong even linguistically, as the second character does not mean opportunity, but rather 'incipient moment' or 'decisive point'. Thus, not necessarily a time for optimism or a good chance of advancement, but certainly a period of change.[24])

If crises really are opportunities, why is it a given who will lose? Because it is. It is (almost) always the poor who pay the price. Crisis as 'danger and opportunity' hides a class character: danger for whom and opportunity for whom? For the ruling class, crises can indeed be opportunities. The famous saying 'never let a good crisis go to waste' (often attributed to Winston Churchill) also comes with a class character. Try saying that to the thousands losing their loved ones in wildfires, heatwaves and floods, to the millions losing their jobs and homes in

23 Cited in Robinson Meyer, 'Al Gore Dreamed Up a Satellite – and It Just Took Its First Picture of Earth', *Atlantic*, 20 July 2015.

24 Victor H. Mair, '"Crisis" Does NOT Equal "Danger" Plus "Opportunity". How a Misunderstanding about Chinese Characters Has Led Many Astray', pinyin.info, September 2009.

economic crises, or to young women and children being forced into prostitution. For workers, the poor and small farmers, especially in poor countries, crises are not opportunities to be 'used'. Crises are desperation, unemployment and death.

Despite the devastating impact of crises on ordinary people, it is not only bourgeois economists and North American presidents who have viewed crises with a degree of hope and optimism. The young Karl Marx was basically preparing for the fall of capitalism as soon as he saw signs of crisis on the horizon.[25] Engels was not much different. In 1845, Engels wrote that the people 'will not endure more than one more crisis'.[26] But the next crisis in 1847, in the midst of the Europe of revolutions, quickly passed. Hope returned with the Great Crisis of 1856–57. Engels wrote to Marx in November 1857: 'Physically, the crisis will do me as much good as a bathe in the sea; I can sense it already. In 1848 we were saying: Now our time is coming, and so in a certain sense it was, but this time it is coming properly; now it's a case of do or die.'[27] Marx was working on the *Grundrisse* at the time and wrote in a letter to Engels that he was working like mad at night to finish the manuscript before the flood came.[28] Regardless, the crisis of 1857 passed without any revolution; there was no 'do or die'. Instead, the crisis was followed by a prolonged economic boom.

The young Marx's optimism did not come out of nowhere, and we can better understand this with a short return to the conceptual history. Milstein argues for a defensive reading of crisis developing during the seventeenth century, which can be linked to Hobbes's *Leviathan* and was about overcoming dangers and restoring a 'normal state'. In contrast, what Milstein calls an 'offensive reading' of crisis developed during the eighteenth century, with writers such as Rousseau and Thomas Paine. This was no longer about retreating or trying to avoid crises, but, rather, about moving on to the next stage of historical development.[29] In this

25 Sven-Eric Liedman, *A World to Win: The Life and Works of Karl Marx* (London: Verso, 2018), ch. 14.

26 Friedrich Engels, *The Condition of the Working Class in England in 1844* (Mansfield: Martino Publishing, 2013), p. 296.

27 Friedrich Engels, 'Engels Letter to Marx, Manchester, 4 August 1856'. Reprinted in Karl Marx and Friedrich Engels, *Collected Works*, vol. 40 (London: Lawrence & Wishart, e-book, 2010).

28 See Liedman, *A World to Win*.

29 Milstein, 'Thinking Politically', pp. 144–5.

respect, the younger Marx is surely a child of the eighteenth century. The older Marx gives us a very different approach to crisis, and according to Peter Thomas and Geert Reuten, the *Grundrisse* is the battleground for the two different perspectives.[30] In sharp contrast to all previous naïve optimism, the older Marx emphasised how crises functioned within phases of accumulation cycles and were components of the reproduction of capital.

Many Marxists never stopped hoping that crises would be opportunities, even with revolutionary potential. Environmental historian Jason Moore argues that, while crises are full of dangers, 'as the Chinese would remind us, they are also full of opportunity'.[31] If any Chinese have actually reminded us of this very point, they have probably studied Western crisis theory. Moore subtitles one of his most famous texts 'How I Learned to Stop Worrying about "The" Environment and Love the Crisis of Capitalism'. It is not clear from the text what this means, but, elsewhere, he has argued that the fall of the Roman Empire after the fourth century and the collapse of feudal power in the fourteenth century led to a golden age in living standards for the vast majority.[32] This might be empirically true, but it remains politically irrelevant to speculate today about positive outcomes centuries into the future. For someone losing their loved ones due to crises, the prophecy that someone else's great-grandchildren's grandchildren might benefit from the current disasters is hardly a reason to learn to love any crisis.

The general tendency throughout the history of capitalism is that crises do not tend to benefit workers and the poor, but there might be exceptions to the rule. One is the Black Death which, although it

30 Peter D. Thomas and Geert Reuten, 'Crisis and the Rate of Profit in Marx's Laboratory', in Riccardo Bellofiore, Guido Starosta and Peter D. Thomas (eds), *Marx's Laboratory, Critical Interpretations of the Grundrisse* (Leiden: Brill, 2013), p. 312.

31 Jason W. Moore, 'Toward a Singular Metabolism: Epistemic Rifts and Environment-Making in the Capitalist World-Ecology', *New Geographies* 6 (2014), p. 16. For other examples, see Dan Cunniah, 'Preface', *International Journal of Labour Research* 1, no. 2 (2010), pp. 5–7; Salar Mohandesi, 'Crisis of a New Type', *Viewpoint Magazine*, 13 May 2020; Jessop, 'The Symptomatology of Crises', p. 246.

32 Jason W. Moore, 'The End of Cheap Nature: Or How I Learned to Stop Worrying about "The" Environment and Love the Crisis of Capitalism', in C. Suter and C. Chase-Dunn (eds), *Structures of the World Political Economy and the Future of Global Conflict and Cooperation* (Berlin: Lit Verlag, 2014), p. 285. Jason W. Moore, *Capitalism in the Web of Life: Ecology and the Accumulation of Capital* (London: Verso, 2015), pp. 86–7.

occurred before capitalism, is still a relevant example. Small farmers and the poor who survived the plague were then in a better position, but at the cost of having lost friends and family in a terrible mass death. Cholera made life terrible in nineteenth-century industrialised cities, but, arguably, contributed to public health measures and urban planning that gave workers a better local environment. Should we coldly ignore social consequences and consider plagues and cholera as opportunities for the working class? At what cost?

Concerning economic crises, the most common example of the working class advancing its position through a crisis is the interwar period. Certainly not everywhere, but in places like Norway, Sweden and the US we must ask: Did the workers' movement win *because of* the crisis? The working class had been strengthening its position and building its movement for years – was this really reinforced by, say, the Great Depression of 1929? These are complicated questions, to which we will return later in the book. Here we just need to emphasise that what we are discussing are possible *exceptions* to the main tendency.

The argument of crisis as opportunity can also be taken a step further. Some feel that it is only through crises that the left can find political opportunities. The 2019 and 2020 elections with Jeremy Corbyn and Bernie Sanders were often described as 'once-in-a-lifetime' opportunities. Perhaps too inspired by Gramsci, some pushed the thesis of crises as decisive 'populist moments' and breaking points between different forms of hegemony: yesterday was too early; tomorrow is too late! Crisis is the only opportunity for real radical change; if we lose *now*, we need to wait forty to fifty years for the next hegemonic crisis. Fortunately for us, this is wrong.

According to the Swedish historian Kjell Östberg, economic crises do not necessarily create rebellion and radicalisation. Social struggle shows a relatively independent relationship with economic cycles and with long as well as short economic waves. If anything, there seems to be a negative correlation between economic crises and higher unemployment, on the one hand, and widespread readiness to fight, on the other.[33] Looking quickly at the twentieth century, we see that

33 Kjell Östberg, 'Den solidariska välfärdsstaten och förändringarna i den politiska dagordningen', in Torsten Kjellgren (ed.), *När skiftet äger rum: Vad händer när den politiska dagordningen ändras* (Stockholm: Tankesmedjan Tiden, 2017), pp. 25–8; Kjell Östberg, *Folk i rörelse: Vår demokratis historia* (Stockholm: Ordfront, 2021), pp. 65, 100, 150.

widespread protests seem to take place a few years *before* the crisis. The 1917 revolution came in a sea of wars and crises but took place twelve years before the great crisis of 1929; the 1968 uprisings came five years before the 1973 crisis; and the anti-globalisation and anti-war movements of 1999–2003 came a few years before 2008. Should we conclude from this that great opportunities always come a few years before major economic crises? No, that would also be far too speculative. Having said that, we should acknowledge that social struggles certainly do not happen independently from political economic processes. But, rather than searching further for such historical relations in this respect, the aim of this book is to help us understand the nature of crises so that we know the terrain on which we will need to fight the coming crises.

Östberg finds it hopeful that waves of radicalisation are not determined by economic waves, as insurgencies are therefore not dependent on specific economic cycles. But this does bring further problems for the crisis-as-opportunity approach: if chances for radical change are *at least* as high during periods not characterised by crisis as they are during crisis, then every single day with or without any crisis is an opportunity. Here, the concept becomes politically and analytically meaningless. The crisis-as-opportunity argument arguably peaked in 2015. Five years after the earthquake in 2010 in Haiti which killed around 230,000 people and left 1.5 million homeless, a writer at the *Correspondent* had the audacity to ask whether the earthquake wasn't also a 'fresh new opportunity'. Perhaps even 'the best thing that ever happened to Haiti?'[34]

Another version of the opportunity thesis is one that sees crisis and progress everywhere. Brian Milstein argues that social welfare institutions and human rights have been established and many ideas of socio-economic justice have become mainstream because of, and in the wake of, economic crises.[35] The problem here is that because capitalism has created so many crises, and since major institutional changes develop over years, it is not hard to find a crisis that took place a few years before or after any important political decision. This does not necessarily mean that the crisis is the *cause* of the improvement.

---

34 Maite Vermeulen, 'Was the Earthquake the Best Thing That Ever Happened to Haiti?', *Correspondent*, 12 January 2015.

35 Milstein, 'Thinking Politically', p. 142.

If crises are indeed opportunities, should we hope for more crises? That would be ridiculous. The idea that crises are good because they open up opportunities for the poorest is just as absurd as the idea that the slave trade opened up opportunities for today's African Americans to become entrepreneurs and even presidents of the United States. Or that colonisation was an opportunity for the poor in India, for example, because it gave them buildings and railways. Only fascists or psychopaths would make such arguments. These are anti-humanist positions that calculate with – or rather ignore – the lives of vulnerable people.

If opportunities – as defined in textbooks – are occasions or situations that make it possible to do something you want or have to do, and if opportunities – as conventionally understood – entail moments of excitement, optimism and hopefulness, and chances for advancement, then we must refrain from referring to crises as opportunities for the working class, the environmental movement or the political left. This does not mean we should not attack crises with all our might. We just need a different approach.

Beyond Keynesianism

Throughout the 2010s, you could go to conferences where Marxists discussed crisis theory and how crises must be solved through revolutions and socialism. Then we all went home to our respective socialist parties and voted for Keynesian investment programmes. Why do socialists run to Keynes every time there is a crisis?

Costas Lapavitsas explicitly says that Keynesianism is the most powerful tool we have, even as Marxists, to deal with political issues in the here and now. While the Marxist tradition, according to Lapavitsas, is good at understanding and dealing with medium- and long-term problems, it cannot be compared to Keynesianism when it comes to short-term crisis management.[36] If Lapavitsas has a point – that Keynesianism is the best tool Marxists have in the face of crises – he has, above all, pointed to a major problem.

But, if crises are mainly possibilities for the ruling class and problems for the rest of us, and if the struggle for socialism would be easier without

36 Costas Lapavitsas, 'Greece: Phase Two. An Interview with Costas Lapavitsas', *Jacobin*, 3 December 2015.

crisis, could we quickly solve crises with Keynesianism and return to Marxism as soon as the storm is over? This is a dead end. Apart from the fact that there is no guarantee that Keynesian crisis management actually solves crises, the crises are so many and so severe that a left that mobilises a social-liberal approach in every crisis will be stuck there.

Keynesian crisis management may to a greater or lesser extent be directed at servicing workers and the poor, but, as with any inter-capitalist solution, it will always have to restore profits and reproduce capitalism. This is a prerequisite. And one that can be easy to forget. With arguments about state interventions, challenges to the power of certain capitalists and calls for grand reforms – add to this that Keynes himself was part of the legendary Bloomsbury Group – Keynesianism can offer a 'critical edge', a sense of radicalism, although it will always save capitalism, one crisis after the other.

It is easy to dismiss Keynesianism as liberal theory masquerading as critical theory. But, as soon as crises become concrete, things become more difficult. There are reasons why socialists so often grasp for Keynesianism in crises. Left-Keynesian approaches do seek to implement social reforms that can improve the lives of workers and the poor. Easing the pain for the working class without confronting the ruling class is, arguably, better than not easing working-class pain at all. If someone needs a crisis to vote for investments in public transport, this is surely better than no such investments at all. For socialists in the face of actual crises, there are seldom better alternatives on the table. Even for Marxists, this tends to be the least bad option. Keynesians may find it hard to admit the big truth: that capitalism itself is the problem. But Marxists find it equally difficult to know what to do with this great truth in the midst of a crisis.

Crises create shocks in situations where people demand political action. There might be much uncertainty in the air, but *something* must be done. The hypothetical alternatives of allowing the economic crisis to deepen or the climate crisis to escalate are, by most people, considered worse than those offered by the powers that be. The gravity of the situation – both how serious the situation is, and how little time there is to respond to it – pushes many to search for safe havens in less radical circles. We can call this the pragmatic trap, or perhaps the Keynesian fishing net: the left is caught between different choices, all of which are calibrated to reorganise capitalism. This is just as true for economic crises as it is for ecological ones. It is in such situations that the climate-conscious left bends its neck and says yes to Hillary Clinton and Joe Biden – because the alternative is Donald Trump.

On the one hand, a left that accepts Keynesianism as crisis policy is a left that keeps capitalism alive, which makes the system ready for new rounds of exploitation, accumulation through dispossession and destruction of nature. Given how often capitalism produces crises, if we do not find another approach, the left will be busy reproducing capitalism for decades and decades to come. On the other hand, a left that cannot handle the here and now of crises, that cannot speak to the social distress that crises produce and that operates only on a discursive level of revolution and smashing the system will forever be politically irrelevant. We still need another approach.

We can neither escape nor ignore the crises. I see no reason to criticise individuals or groups who try to escape capitalism, either by living 'outside' the system within urban centres or by moving to the countryside or into the wild. But the vast majority of workers will still be left in the coils of the crises of capitalism. As the crises of capitalism are global, they cannot merely be confronted at a local scale. There is nothing wrong with deep ecologists moving to the country and growing their own food, but this type of response will not solve the major problems in a world of 8 billion people. Local mutual aid responses to crises might ease some pain during a crisis and create community solidarity. There are many reasons to support, and indeed participate in, this type of response. But socialists must also look a few steps further. It is not only about surviving the crises; it is about stopping them.

Then there is Naomi Klein, who emphasises the need to remain calm in the face of shocks and avoid being carried away by panic.[37] This might be wise advice for some pseudo-crises or in the face of conspiracy theories. However, crises are not only discourses; they are actually existing events that shake the world. The shock is real. When people see their jobs, housing and the earth beneath them disappearing quickly, the strategy of organising the masses to keep calm will hardly win. I have a softer spot, in this respect, for Greta Thunberg's 'act as if the house was on fire, because it is'.[38] We need to 'panic together', and we need

37 Naomi Klein, *Doppelganger: A Trip into the Mirror World* (London: Allen Lane, 2023), ch. 11.

38 Greta Thunberg, 'Our House Is On Fire', *Guardian*, 25 January 2019.

organised socialist movements that bring our own shock doctrines and creative destructions into the ring.

A position that is very rare in Marxism is to try to ignore or disregard crises altogether. One exception was the Italian Communist Party in 1975, which declared that there was no need to dramatise the crises because they obscured the true state of affairs and made it more difficult to find solutions.[39] This never proved a very productive strategy. When the crises *are* the state of affairs, we need to face the challenge: we must confront the crisis.

Towards a Socialist Approach

According to the Marxist economist Rikard Štajner, there are two cataclysms of mankind: war and crisis.[40] What Štajner is indicating is that we should relate to capitalist crises in the same ways that we approach war, hunger, slavery and so on. This approach I believe is fruitful. Crisis and its causes are something we must *fight against*. Rather than opportunities we look forward to exploring, or moments when the fight for socialism is put on hold, the crises are problems we must solve. Štajner's linking of war and crisis is also interesting from a historical perspective. In the 1910s, the struggle for revolution was not just a battle between workers and capitalists in workplaces. It was also crucial to ending (or preventing) capitalist/imperialist wars. Socialism in our time must be about stopping the crisis. Rather than hope and excitement, socialists should approach the crises of capitalism with rage and anger. Rather than opportunities, crises are the enemy.

Lenin said that war is not something you can end 'at will'; similarly, crises are not something we can choose to pause under capitalism.[41] Stopping crises requires something more radical than a few regulations or a more active state. Over a hundred years ago, those who opposed war sought to expose its class nature: Who was sacrificed and who supported the war; what interests did it serve; what historical and economic

39 Rikard Štajner, *Crisis: Anatomy of Contemporary Crises and (a) Theory of Crises in the Neo-imperialist Stage of Capitalism* (Belgrade: KOMUNIST, 1976), pp. 66–7.

40 Ibid., p. 190.

41 V. I. Lenin, 'The Tasks of the Proletariat in Our Revolution', in *Lenin Collected Works*, vol. 24 (Moscow: Progress Publishers, 1964 [1917]), pp. 55–92.

conditions produced it, and how did wars reproduce capitalism? In a similar way, we must expose the role of crises under capitalism.

A socialist approach to crises cannot be based on any naïve optimism that crises are 'opportunities', or sweet dreams that crises will provide us with the collapse of capitalism. We must start from what normally happens during actually existing crises, and an understanding of how *capitalism produces crises* and *crisis reproduces capitalism*. In this book, we will see that it is empirically far-fetched to call the crises of capitalism opportunities for the working class or the political left, and we will discuss theoretically how this can be the case. The crises of capitalism are not moments of truth; they are battlefields. There are reasons why (parts of) the ruling class – not workers and the poor – tend to win these battles, and, in order to do something about this, we must identify the reasons. Therefore, we will in this book also examine creative destruction, the class character of crisis, crisis as shocks and panic, the relative autonomy of the state, and the role of nationalism, racism, fascism and war. And more.

That the crises of capitalism are social paroxysms means that they necessarily exist on different levels. So, then, must any socialist approach that seeks to confront the crises. On a general level, we must understand the nature of crises, how crisis produces capitalism and vice versa. We can call this a Marxist *crisis critique*. Once we know the terrain, we can start articulating more concrete socialist *crisis policies*, which are general strategies and programmes that socialists can use to confront actual crises. But, when a crisis hits, theoretical understandings and general programmes are insufficient. There is an urgent need for very concrete action. Socialist *crisis management* is needed to ease social pain for the working class and to bring the class character we prefer directly into situations of shocks and panic.

The aim of this book – standing on the shoulders of giants, in dialogue with comrades – is to explore what a socialist approach to crisis can look like. The scope is limited to crisis critique, with only brief discussions about crisis policy towards the end. This means that much more work needs to be done. I hope that some readers will feel a calling.

History has shown that crises are not usually opportunities for workers and the poor; but there is no reason to bend the stick too far in the other direction. This is not an iron law. It is a tendency. Our historical mission as socialists in a burning world is to make a monumental exception to this tendency.

1

When the Economy Is Not Healthy . . .

Economic crises existed before capitalism and will in all likelihood still occur after capitalism. Yet, economic crisis under capitalism has a distinct nature. Economic crises before capitalism – to put it bluntly – normally occurred because of external shocks, be it wars, floods, droughts and earthquakes, plagues or other epidemics. In contrast, economic crises under capitalism are, following the Hungarian Marxist Georg Lukács, merely an increase in the degree and intensity of everyday life in bourgeois society.[1] While pre-capitalist economic crises were often closely linked to people having too little of various things, what we might call an underproduction of use-values, crises under capitalism are often associated with an oversupply of goods, food and labour – an overproduction of exchange-values.[2] In pre-capitalist crises, material destruction was a cause of crisis; in capitalism, on the other hand, such destruction is a *result* of crisis. Ironically, destruction has become part of the solution to the crisis. Before capitalism a famine, for example, would be characterised as an economic crisis, because it made it difficult for people to reproduce their social formation. Under capitalism, it is by no means a given that famine is considered an economic

1 Georg Lukács, *History and Class Consciousness: Studies in Marxist Dialectics* (Pontypool: Merlin Press, 1971), p. 101.

2 E.g., Paul Sweezy, *The Theory of Capitalist Development* (London: Dennis Dobson, 1946), ch. 8; Ernest Mandel, *The Second Slump: A Marxist Analysis of Recession in the Seventies* (London: NLB, 1978), p. 167.

crisis. Capitalism can function perfectly well even if many people starve to death.

Economic Crisis

If we focus exclusively on the panic that arises in stock exchanges and markets, we can see interesting traces even to events that occurred long before capitalism: the first bank crash is said to have occurred in ancient Athens in 354 BC, and the first bank run in the 1340s when poor returns on silver handling in Ragusa (Dubrovnik) caused a rush to a bank in Venice that had invested heavily in silver. From the long birth of capitalism, the Amsterdam Tulip Crash of 1636–37 is the most legendary crisis. In a matter of weeks, prices multiplied in what became known as the world's first speculative bubble. The price of the Switzer – one of the more popular tulip bulbs – rose by no less than 1,200 per cent between 31 December 1636 and 3 February 1637.[3] Then things turned. The crash has often been attributed to mass psychosis and folly. Already, we are able to see that the madness appears within a system: what made the madness possible was that people could buy options on bulbs with money they did not have, and at a scale and pace that had hardly been possible before.

Not only are the crises of the sixteenth, seventeenth and eighteenth centuries interesting in their own right, but several aspects are relevant to discuss even today. Such as that euphoria, mania and panic create unrest; that state and capital interact; and that war often plays a key role in crises. The effects of the crises of early capitalism were usually relatively limited. Crises like the Tulip Bubble, or the Mississippi and South Sea Bubbles less than a century later, did create panic and shock, but because they did not occur within a developed capitalist economy, the impact was limited. This is in stark contrast to what was to come.

What we have come to know today as economic crises occur only when capitalism is relatively well developed as an institutional system. By this, I

3 Lars Magnusson, *Finanskrascher: från kapitalismens födelse till Lehman Brothers* (Stockholm: Natur & Kultur, 2020), pp. 46–7, 64, 81; Charles P. Kindleberger, *Manias, Panics, and Crashes: A History of Financial Crises* (New York: John Wiley and Sons, Inc., 2000), pp. 109–11.

mean a system where goods and services are produced for profit; where private ownership of the means of production is the central form of ownership, and capitalists versus wage labourers becomes the dominant class relation; where surplus value comes from the exploitation of wage labour; where capital becomes the keynote of the economy's melody; where firms must constantly find innovative improvements and new technologies because of competition; where banks are part of international systems; where large investments are often made on credit; where workers become consumers of goods on a large scale; and where the main rule is that production is for a market – not for direct exchange or for own consumption.

Capitalism is to be understood here as the institutionalised and historical social system – or perhaps rather the ruling order – that enables capital accumulation to function as the very lifeblood of the political economy. Prior to capitalism, capital existed, in Mandel's formulation, on the *outside* of the dominant production processes, whereas under capitalism, capital dominates the main sectors of production. Although capitalism developed gradually from the sixteenth century onwards, it took a few centuries to find its modern form and its proper name.[4] It is only with the Industrial Revolution that we can begin to talk about a truly developed capitalism, and economic crises take on a different character thereafter.

In the nineteenth century, crises began to flourish. They occurred with a regularity and intensity that shocked the world and shook the core countries of capitalism: 1825, 1837, 1847, 1857, 1866, 1873, 1882, 1891.[5] Not only did the use of the term 'crisis' become more common, but explanations were needed for why they occurred. Until the mid-nineteenth century, it was common to explain the crises solely in terms of extraordinary events. The crisis of 1825 took the form of a financial crash in Britain, largely involving speculation in South American securities (with large

4 Ernest Mandel, 'Introduction', in Karl Marx, *Capital*, vol. 2 (London: Penguin, 1978), p. 18. See also Henryk Grossman, 'Marx, Classical Economics, and the Problem of Dynamics', *International Journal of Political Economy* 36, no. 2 (Summer 2007), p. 59; Karl Marx, *The Poverty of Philosophy* (Paris: Progress Publishers, 1955), pp. 8, 28; Kevin B. Anderson, *Marx at the Margins* (Chicago: University of Chicago Press, 2010), p. 188.

5 Mandel, *The Second Slump*, p. 34; Rikard Štajner, *Crisis: Anatomy of Contemporary Crises and (a) Theory of Crises in the Neo-imperialist Stage of Capitalism* (Belgrade: KOMUNIST, 1976), p. 165.

investments in a country in South America that did not even exist); in Germany that same year, the crisis was related to agricultural surpluses.[6] The 'Panic of 1837' in the US was explained by the speculative lending practices in the Wild West, falling cotton prices and the fact that the Bank of England raised interest rates to compensate for increased imports of agricultural products. And there was a gigantic real estate bubble in the midst of it all. A decade later (1847–48) came a crisis associated with railway construction and poor harvests (in Germany). Another decade later (1856–57) came the short but sharp crash associated with the gold rush and the mad speculation it brought, but also with speculation in building and railways (again). The crisis of 1866 is often associated with the US Civil War and the sudden shortage of cotton. The 1873 crisis began with the collapse of the Vienna stock exchange linked to the real estate market and was triggered in the US by speculation in railways (again!) and exacerbated by war between Germany and France.[7]

As the crises seemed to be strung together like beads on a necklace, people began showing an interest in examining not only the beads but also the thread that tied them together. On the surface, nineteenth-century crises often appeared to be different, although railways, real estate and credit were regularly recurring themes. Were there any underlying processes linking the crises together? Rather than one trade crisis after another as independent phenomena, both socialists and liberals sought to understand them as 'crises of production', caused by the capitalist system itself.

Discussions on the causes of the crises gave rise to some explanations that, despite their weaknesses, are still used today: human greed, the phenomenon of money and technological innovation. But human nature and the phenomenon of money are much older than capitalism, and technological change is more a consequence of the constant need

6 Reinhart Koselleck, 'Crisis', *Journal of the History of Ideas* 67, no. 2 (2006), p. 389; Rosa Luxemburg, *The Complete Works of Rosa Luxemburg*, vol. 2, *Economic Writings 2*, ed. Peter Hudis and Paul Le Blanc (London: Verso, 2015), ch. 30.

7 On the nineteenth century: for an overview see Magnusson, *Finanskrascher*, chs 6–7; for a conceptual perspective see Koselleck, 'Crisis', pp. 365–6, 381–97; on speculation see Kindleberger, *Manias, Panics, and Crashes*, ch. 8; on export as crisis management see, e.g., Luxemburg, *The Complete Works*, vol. 2, chs 30–1; on long waves see Ernest Mandel, *Late Capitalism* (London: Verso, 1978), ch. 4; Michael Roberts, *The Long Depression: How It Happened, Why It Happened, and What Happens Next* (Chicago: Haymarket, 2016), ch. 1.

for profit than a cause of its fundamental development. Marx would argue that the real obstacle to capitalist production is capital itself, and Marxists would later agree that crises grew out of contradictions in the capitalist system.[8] But it was a little harder to agree on exactly how.

A Century of Crises: In Theory and Reality

Karl Marx died in 1883, in the midst of what is often called the Long Depression (1873–90) and the Second Industrial Revolution (1871–1914). Marx lived through some turbulent decades, but things would only escalate after his death: colonisation, more recurring crises (1900, 1907, 1913, 1921), two world wars, socialist revolutions, an emerging fascism, and the Great Crash of 1929. No wonder the decades after the turn of the century were a golden age for Marxist crisis theory.[9]

The theory of overproduction became the first dominant theory of crisis after Marx's death, further developed by Friedrich Engels and Karl Kautsky. Briefly, the theory goes like this: We must remember that the major consumer group under capitalism consists in the wage-earners. Since a certain proportion of a firm's turnover must go to profit, there is an inherent problem in the system: wage-earners can never receive high enough wages to buy all the goods they have produced. One could imagine that firms would meet such a hurdle by producing less or by getting people, somehow, to consume more, so that equilibrium is restored. Because of competition and the anarchy of the market, no one can shrink production and see their profits fall. Capitalists must meet the challenge

8 See, e.g., Karl Marx, *Capital*, vol. 3 (London: Penguin, 1981), p. 358; Daniel Bensaïd, 'The Time of Crises (and Cherries)', *Historical Materialism* 24, no. 4 (2016), p. 24.

9 For introductions to Marxist theories on economic crises,, see, e.g., Mandel, *Late Capitalism*; Anwar Shaikh, 'An Introduction to the History of Crisis Theories', in Union for Radical Political Economics (ed.), *U.S. Capitalism in Crisis* (New York: Union for Radical Political Economics, 1978), pp. 219–40; Štajner, *Crisis*; James O'Connor, *The Meaning of Crisis: A Theoretical Introduction* (Oxford: Basil Blackwell, 1987); Sweezy, *The Theory of Capitalist Development*; Chris Harman, *Explaining the Crisis: A Marxist Re-appraisal* (London: Bookmarks, 1992); Simon Clarke, *Marx's Theory of Crisis* (London: St Martin's, 1994); David Harvey, *The Limits to Capital* (London: Verso, 1999); Bertell Ollman, *Dance of the Dialectic: Steps in Marx's Method* (Chicago: University of Illinois Press, 2003); Radhika Desai, 'Consumption Demand in Marx and in the Current Crisis', *Research in Political Economy* 26 (2010), pp. 101–43.

by producing more efficiently, cutting wages, implementing new technologies and simply producing more, which immediately benefits individual capitalists, but exacerbates the general problem of overproduction.

For Marx, the theory of overproduction was a critique of the French liberal economist Jean-Baptiste Say. Say argued that since every sale is also a purchase, supply and demand would, in principle, always be the same. Production thus created its own demand.[10] But capitalism does not produce for us to consume – it produces for profit. And the exchange of commodities is mediated by money. So, while capitalists sitting on some physical commodities may be absolutely desperate to sell, those with money do not have to buy. As money is a general means of payment, one can sit on the money and wait. The fact that money divides trade into two acts has therefore been called an 'embryo of crises'.[11]

There is, thus, a contradiction between the limited consuming capacity of the masses and the development of the productive forces which do not take into account the limits of the market. Production necessarily runs away, ahead of the capacity of the market to swallow up all commodities. And this, according to the theory, ends in crisis. This is an absurd aspect of capitalism, well put by Kautsky: 'Thousands perish from cold and hunger because they have produced too much clothing, too much food, and too many houses!'[12]

Kautsky's theory of overproduction would be criticised by Eduard Bernstein, among others. Bernstein argued that the risk of overproduction was small because it was counteracted by three things: the growth of the domestic market, an expanding middle class and the emergence of foreign markets through imperialism. Kautsky responded to the criticism by approaching what would become the next dominant theory in Marxism: underconsumption.[13]

Although underconsumption is often mentioned as a separate theory alongside the theory of overproduction, it can be difficult to understand the difference between them. Some theorists, such as

10 Lars Pålsson Syll, *De ekonomiska teoriernas historia* (Lund: Studentlitteratur, 2007), pp. 137–9.

11 Karl Marx, *Grundrisse* (New York: Vintage, 1973), p. 198; Bensaïd, 'The Time of Crises', pp. 21–2.

12 Karl Kautsky, *The Class Struggle (Erfurt Program)* (Chicago: C. H. Kerr, 1910), p. 80.

13 Clarke, *Marx's Theory*.

Derek Wall, see the two as essentially the same theory. Anwar Shaikh and Simon Clarke argue the difference lies in the fact that the traditional view of underconsumption theory defines the demand for consumer goods as what regulates total production. One criticism of underconsumption theory is that, because it focuses mainly on consumer goods, the theory tends to miss what we might call productive consumption – that is, that industries and factories need to consume means of production, such as machinery. Štajner further stresses that the crises of capitalism cannot only come from insufficient consumption, as this also occurred in pre-capitalist times but without creating similar crises.[14] Nonetheless, the theory of underconsumption gained ground in the East and the West. In the US, underconsumption became an important component of Paul Sweezy's and others' newly developed stagnation theory.[15] Through the work of the Hungarian economist Eugen Varga, underconsumption was elevated in Moscow to the official theory of Stalinism.[16]

It seems to me that the differentiation between overproduction and underconsumption has also changed over time. In the 1970s, realisation crises were often defined as underconsumption. Neither Shaikh, in his introduction to crisis theories from 1978, nor the *Dictionary of Marxist Thought* from 1983, for example, mention overproduction; they only mention underconsumption. Today, the situation seems to have reversed.[17]

The leading alternative to underconsumption theory before the First World War focused on the phenomenon of disproportionality, with the Ukrainian economist Mikhail Tugan-Baranowsky as main protagonist. According to this theory, capital accumulation requires appropriate

14 Štajner, *Crisis*, p. 43; Derek Wall, *Economics after Capitalism: A Guide to the Ruins and Road to the Future* (London: Pluto Press, 2015); Clarke, *Marx's Theory*; Shaikh, 'An Introduction'; Ernest Mandel, 'Introduction', in Marx, *Capital*, vol. 3, p. 44.

15 See, e.g., Paul Baran and Paul Sweezy, *Monopoly Capital: An Essay on the American Economic and Social Order* (New York: Monthly Review Press, 1966). For critique concerning the role of competition, see Anwar Shaikh, *Capitalism: Real Competition, Turbulent Dynamics and Global Crises* (New York: Oxford University Press, 2016); and Brett Christophers, *The Great Leveler: Capitalism and Competition in the Court of Law* (Cambridge, MA: Harvard University Press, 2016).

16 Rick Kuhn, 'Henryk Grossman and the Recovery of Marxism', *Historical Materialism* 13, no. 3 (2005), p. 90; Clarke, *Marx's Theory*, p. 8.

17 Shaikh, 'An Introduction'; Tom Bottomore, Laurence Harris, V. G. Kiernan and Ralph Miliband (eds), *A Dictionary of Marxist Thought* (Oxford: Blackwell, 1983).

proportional relationships between different branches of production; there may be overproduction in some industries and underproduction in others. A crisis occurs first in a particular branch of production where overproduction reaches its limit, whereupon the collapse risks generating chain reactions in other branches. Imbalances could be kept in check, at least for some time, by the credit and banking systems.

Where overinvestment was considered rational by Engels and Kautsky, because capitalism is based on competition that does not take into account the limits of the market, for proponents of disproportionality such as the German Austrian economist and politician Rudolf Hilferding, overinvestment was the result of capitalist misjudgements. A political implication of such a theory was that, through comprehensive planning, the tendency to crisis could be overcome.[18] The theory thus tended to fall into corporatist conclusions and also became associated with the fascist corporation (without Hilferding in any way going over to the dark side; on the contrary, he was killed by the Gestapo in 1941).

Disproportionality never became a dominant theory. The theory was criticised for its one-sided focus on capitalists and lack of interest in the relations between labour and capital. Within the communist movement, underconsumption and the plight of the poor working class were still the focus of the interwar period. The underconsumptionists accused disproportionality theory of being a reformist aberration. Ironically, in the post-war period, the arguments that raising wages and increasing government spending could provide a renewed stimulus to accumulation became so dominant that it was now underconsumptionism that would be labelled social democratic and reformist. Before discussing this further, let us first turn to capitalism's first existential crisis.

Capitalism: Doomed and Dying?

After the turbulent beginning of the twentieth century, capitalism began to take off again in the 1920s. After the recession of 1920–21, many

18 Clarke, *Marx's Theory*, p. 31; Mandel, *Late Capitalism*, pp. 34–5. For an early critique of the thesis that crises within capitalism could be eliminated through planning, see V. I. Lenin, *Imperialism, the Highest Stage of Capitalism*, in *Selected Works* (Moscow: Progress Publishers, 1963), available at marxists.org.

countries experienced an uninterrupted boom throughout the decade, to a great extent – not least in the United States – based on the production and consumption of cars, household goods, chemicals and housing. An industrial sector pumped out a torrent of new goods, and a consumer sector ran on credit. The assembly line principle was developed in production, while instalment purchases became more common, and stocks and real estate became a hobby for a growing portion of the population. From this era comes the concept of the Ponzi scheme – named after Carlo Ponzi, the man who promised that anyone who invested $10 with him would get $30 back within sixty days. Everyone wanted to become rich. So, in 1929, the balloon burst.[19]

This was not only considered a crisis of capitalism. It was the first time an economic crisis made people think that capitalism itself might be in trouble. Commentators on both the right and the left thought they were witnessing the demise of capitalism in real time, and the *New York Times* captured this view with its headline on 5 March 1934: 'Capitalism Is Doomed, Dying or Dead'.[20] Until then, the dominant notion had been that crises would resolve themselves without government intervention, that the free capitalist economy had a capacity of its own to almost automatically cure its own ills through an adjustment in supply and demand.[21]

Although laissez-faire was on the defensive – in 1925, Keynes had already written of 'the end of the laissez-faire system' – the original policy in the US in 1929 was, according to the North American economist John Kenneth Galbraith, still 'a practical expression of *laissez-faire*'.[22] This only compounded the problems. When the crisis had shaken the economic, social and political structures to their foundations, and with socialist revolutions actually on the agenda, the ruling class no longer dared to rely on any self-regulating mechanisms or free markets.[23]

19 John Kenneth Galbraith, *The Great Crash 1929* (London: Penguin, 2009), chs 2–5.

20 Matt Huber, *Lifeblood: Oil, Freedom, and the Forces of Capital* (Minneapolis: University of Minnesota Press, 2013), p. 33; Štajner, *Crisis*, p. 206.

21 Štajner, *Crisis*, p. 56.

22 Galbraith, *The Great Crash 1929*, p. 160, italics in original; John Maynard Keynes, *Essays in Persuasion* (New York: Classic House Books, 2009), pp. 170–5.

23 On the US, see, e.g., Steve Fraser, 'American Labour and the Great Depression', *International Journal of Labour Research* 2, no. 1 (2010), pp. 9–24; David McNally, *Global Slump: The Economics and Politics of Crisis and Resistance* (Oakland: PM Press, 2011), ch. 3.

States had, of course, intervened in crises before, but only now did they begin to pursue a deliberate crisis policy to save capitalism. Ever since the stock market crash of 1929, or since the Great Depression, or the Bretton Woods agreements, or since the Second World War, or since John Maynard Keynes – depending on how you read history – economic crises have always ended up on the politicians' table. If the first significant breaking point in the history of economic crises was when capitalism had evolved to a level where recurrent crises began to emerge regularly from the system's own contradictions, we have now reached the second breaking point: the capitalist state apparatuses became the guarantors of the survival of the capitalist system at every economic crisis.

The ruling class needed active states to prevent crisis and revolution, and then they found Keynes. That Keynes became the most important economist of his time was not easy to accept for Joseph Schumpeter, the Austrian economist of the same age. People sympathised with Keynes's theories, Schumpeter suggested, not because of the theories themselves, but because of his political-economic recommendations.[24] This is an interesting assertion. Keynes's theories may be complex and contain elements of social analysis and psychology, and there are certainly different ways of reading him, but the political implications were and are straightforward. The anti-cyclical crisis policy is so simple that most politicians can keep up. If the economy is doing well and there is a danger of inflation, the government should hold back, cut spending, and raise taxes and the repo rate. If the economy is doing poorly, the government should increase public spending, rev up the money-printing presses or cut taxes and interest rates.[25] Through the Keynesian multiplier, chain reactions will then create much larger effects in the economy than the actual efforts of the government. If aggregate demand increases,

24 Thomas K. McCraw, *Prophet of Innovation: Joseph Schumpeter and Creative Destruction* (Cambridge, MA: Harvard University Press, 2007).

25 See, e.g., Shaikh, *Capitalism*, ch. 12; John Maynard Keynes, *The General Theory of Employment, Interest and Money* (London: First Harvest, 1964); Paul Mattick Sr, *Marx and Keynes: The Limits of the Mixed Economy* (London: Merlin, 1971); Robert Skidelsky, *Keynes: The Return of a Master* (New York: Public Affairs, 2009); Geoff Mann, *In the Long Run We Are All Dead: Keynesianism, Political Economy, and Revolution* (London: Verso, 2017); Štajner, *Crisis*, pp. 56–9; Paul Krugman, *End This Depression Now!* (London: W. W. Norton, 2012). For Keynesianism as hegemony, see Stuart Hall et al., *Policing the Crisis: Mugging, the State and Law and Order* (London: Macmillan, 1978), pp. 212–14.

the whole economy will be pulled out of the crisis and confidence in the economy will be restored. The theory gave new life to the liberal dream. Capitalism could be achieved without deeper crises, but only if the state actively regulated and stabilised the economy. And it seemed to work.

'Oil Crisis'

The post-war period has often been called the golden age of capitalism. Economic growth was very strong, economic crises were few and the IMF and World Bank were created to regulate away future depressions. Somehow, the theory of underconsumption was turned on its head: instead of explaining stagnation and the inevitability of crises, the insights from the theory – within a Keynesian framework and through an active state – were now used to explain how public consumption and higher wages could create economic stability. Had the liberal dream of a capitalism without crises finally become a reality?

Crises are rude; they do not ask permission before they visit. And, in the early 1970s, crisis struck again. The dominant explanation this time was that the crisis was triggered by OPEC countries refusing to export oil to states that had helped Israel in the 1973 October War. This shocked oil prices, creating inflation and what came to be known as the oil crisis. It is a 'persistent myth', to borrow the term from Mandel, that this was an oil crisis. Such a notion hides not only the underlying processes but also the fact that inflation and economic decline had started earlier.[26]

In the 1970s we saw for the second time a deep crisis in crisis management. The first time, laissez-faire gave way to Keynesianism. By the 1970s, the solution of the 1930s – Keynesianism – had become part of the problem. The Keynesian assumption that the economy would be in either stagnation or inflation was questioned as the crisis now came in the form of *stagflation*: both stagnation and inflation. If one injected

26 Mandel, *The Second Slump*, p. 34, see also p. 22. It has been argued that the crisis first broke out in the real estate sector or the car industry (see Štajner, *Crisis*, pp. 117–26; David Harvey, *A Companion to Marx's Capital* (London: Verso, 2010); Mandel, *The Second Slump*, p. 169), or that it was a wartime inflation caused by the Vietnam War, or 'fiscal liberalisation' and rising wages. On Keynesian crisis management contributing to further inflation, see McNally, *Global Slump*, pp. 30–3. On the oil price shock, see Lennart Schön, *En modern svensk ekonomisk historia: tillväxt och omvandling under två sekel* (Stockholm: SNS Förlag, 2000), p. 436; Huber, *Lifeblood*, ch. 4; Timothy Mitchell, *Carbon Democracy: Political Power in the Age of Oil* (London: Verso, 2013), ch. 7.

more money into the economy, inflation increased; if one did *not* inject more money, stagnation increased. The dominant crisis management had contradictory side-effects. The OECD's advice for 1975 was therefore neither fish nor fowl: stimulate the economy, but gently, in combination with efforts to control inflation.[27] However, the 1973 crisis was not a clear break from Keynesianism.[28] Throughout the 1970s, variants of Keynesianism continued in many quarters; critical discussions on growth and the environment gained momentum while neoliberal ideologies and the political right gained ground. Then came 1979 with two shocks: a new oil shock following the revolution in Iran and the arrival on the scene of Margaret Thatcher in Britain.

The crisis of the 1970s forced Marxists to re-evaluate their theories. Previously, falling profits had been seen as a consequence of overproduction or underconsumption. Now, the fall in the rate of profit seemed to be the very cause of the crisis. When Marxists had to examine the underlying processes of capitalism that had created the crisis, three theories became dominant. One was a variant of overproduction, as discussed previously, while two new theories put profit at the centre. These were the *profit squeeze* and the *law of the tendency of the rate of profit to fall*.

The *profit squeeze* theory said that workers had earned such high wages and expanded rights that it was to the detriment of profits. Lower profits led to lower investment and lower growth, which in turn led to unemployment and stagnation. Left-wing groups used the theory to show that capitalism could not meet the material demands of the working class. Thus, if the working class progressively improved its positions, it would reach a point where it could not be better off under capitalism. However, the theory could also be used by the right: the crisis had been caused by excessive wages, which in turn were driven by an irresponsible radical minority of the working class dominated by undemocratic trade unions. The political implication was exactly the opposite of what the left preached: reduce wages and crush the organised working class.[29]

27 Štajner, *Crisis*, pp. 20, 143–50.

28 Jenny Andersson, 'The Future of the Western World: The OECD and the Interfutures Project', *Journal of Global History* 14, no. 1 (2019), p. 142. On Sweden, see Ståle Holgersen, *Staden och kapitalet. Malmö i krisernas tid* (Gothenburg: Daidalos, 2017).

29 For critique of the profit squeeze, see Bill Dunn, 'Marxist Crisis Theory and the Need to Explain Both Sides of Capitalism's Cyclicity', *Rethinking Marxism* 23, no. 4

The second popular theory put forward to explain the crisis was the *tendency of the rate of profit to fall* (henceforth TRPF). This theory was well known and popular even earlier than the 1970s, but rather than seen as a crisis theory in its own right, it was considered to be a complement to other crisis theories.[30] TRPF must be understood in the tension between two phenomena. First, it is based on Marx's labour theory of value, which says that you can make a lot of money from financial speculation, theft or fraud, but this does not create more value in the system – it just moves money around or creates bubbles. It is only through exploiting workers that *more* value is produced. Secondly, as companies compete, they must constantly find new and more efficient ways to produce. Often, this means investing more in machinery, factories, buildings and the like (called constant capital) than in labour (variable capital). Put simply, machinery will thus constitute an ever-growing share of the economy at the expense of labour. Marx calls this trend an increase in the organic composition of capital.[31] Now the two phenomena meet: as the economy grows, the share of labour compared to the share of machines will decrease, thus reducing the share of surplus value in the system. Even when the economy is growing, the general rate of profit might fall, and the economy will become more and more prone to crisis.

The theory may sound counterintuitive because it is exactly the same processes that first create higher profits for some firms that later lower profits in the economy as a whole. The companies that are most efficient and innovative can indeed increase their profits. Since this improvement will eventually spread to all other companies, it will lead to a levelling out. When all companies have replaced labour with machines – to put it simply – the overall profit rate is squeezed.

Given that capitalism has existed for quite some time despite the fact that the rate of profit tends to fall, an obvious counter-question is why it has not already collapsed. The answer, according to Marx, is that there are a number of counteracting tendencies that can delay,

(2011), pp. 529–31. See also Clarke, *Marx's Theory*, pp. 63–71; Mandel, *The Second Slump*, ch. 5.

30 Cf. Clarke, *Marx's Theory*; Kuhn, 'Henryk Grossman', p. 83.

31 On differences between organic composition, the composition of value and technical composition, see, e.g., Harvey, *The Limits*. Here we work with organic composition as a general concept.

prevent and occasionally even reverse this decline.[32] That there are counter-tendencies should not surprise us. In fact, all tendencies have counter-tendencies. In the case of overproduction, higher wages or increased debt are counter-tendencies, although they are rarely articulated as such.

The most important corrective to TRPF is the crises themselves. When capital is devalued and destroyed on a large scale – and when new sectors with higher profits can replace old ones – the rate of profit gets a fresh start. Other countervailing tendencies are to exploit workers more harshly, to gain access to cheaper fixed capital, or to open new markets – preferably in not-yet-capitalist areas or sectors where the rate of profit is still high, and so on. Countervailing tendencies can slow down or reverse the tendency as such, but when the countervailing factors are no longer strong enough, capitalism becomes unhealthy and crisis-prone.[33]

TRPF has certainly caused some heated debates. Bertell Ollman argues this is largely because people are simply talking about different aspects of the theory. Andrew Kliman shows how the rate of profit can be measured in different ways: Should one measure pre-tax or post-tax, rate of profit, real or expected profit, profit from all sectors or only productive or non-financial sectors, and can one measure within a country or must the analysis be global? The fact that Marx himself used different definitions and measured in different ways does not make things any less complicated.[34] And there is uncertainty about whether TRPF is a direct or indirect cause of economic crises. Could or could not TRPF be the cause of a crisis even if the rate of profit does not actually fall sharply just before a crisis strikes?

Critics argue that if you just cut wages or extend the working day, then, in theory, there will be no crisis. Obviously, there are some limits here. Another objection is that capitalists would not introduce new technology if it leads to declining profits. Such a criticism does not understand the logic of capitalism, where firms *must* constantly find

32 Marx, *Capital*, vol. 3, p. 346.

33 For descriptions of the theory, see Shaikh, 'An Introduction'; Michael Roberts, *The Great Recession: Profit Cycles, Economic Crisis: A Marxist View* (London: Michael Roberts, 2009); Andrew Kliman, *The Failure of Capitalist Production: Underlying Causes of the Great Recession* (London: Pluto Press, 2012); Guglielmo Carchedi, *Behind the Crisis: Marx's Dialectics of Value and Knowledge* (Chicago: Haymarket, 2012).

34 Ollman, *Dance of the Dialectic*; Kliman, *The Failure of Capitalist Production.*

more efficient ways to compete with their rivals. If the countervailing tendencies are as strong as the main tendency, will there, then, ever be crisis? This is, obviously, a question that needs to be addressed empirically. A final objection says that it is always the mass of profit – not the rate of profit – that determines whether capitalists will invest or not. This is partly true, but what is most interesting is the interaction between the mass and the rate of profit.

Neoliberalism and Mass Production of Crises

The concern for crisis theory withered during the post-war decades, even among Marxists. During neoliberalism, in sharp contrast, crises returned en masse, and theorists became again occupied with trying to understand them. In the wake of the 1973 and 1979 oil crises, and the Volcker shock (when US Federal Reserve chairman Paul Volcker in 1981 shocked interest rates to 21 per cent), many poorer countries were plunged into so-called major debt crises because they had borrowed money from the US. Mexico, for example, was close to sovereign default in 1982. The 1990s began with the fall of the Soviet Union. Then capitalism plunged straight into a decade of crises. Scandinavia began the decade with a crisis. Japan entered a crisis in 1991 from which many argue it has still not really recovered. In 1995, both Mexico and Argentina went from euphoria to horror as banks failed and unemployment soared. Then came Southeast Asia's turn. Many Southeast Asian countries had been hailed for their growth and offered seemingly safe investments for Western capital. Then – boom! – panic in the financial markets and the crisis hit there too. By the end of the 1990s, China was almost the only country in the region not in crisis. Then, crises returned to Latin America, above all to Brazil and Argentina – now under the name 'the Asian contagion'.[35]

And, finally, what about Russia? Between the fall of the Soviet Union and 1998, Russia had gained not only the freedom of capitalism, but also a 50 per cent drop in GDP and a halving of income. An estimated

---

35 On the causes of the Asian crises, see Paul Burkett and Martin Hart-Landsberg, 'Crisis and Recovery in East Asia: The Limits of Capitalist Development', *Historical Materialism* 8, no. 1 (2001), pp. 3–47. On the consequences, see M. Ramesh, 'Economic Crisis and Its Social Impacts', *Global Social Policy* 9, no. 1 (2009), pp. 85–92.

225,000 state-owned enterprises were privatised, and average life expectancy fell from sixty-three to fifty-eight years between 1990 and 2000.[36] As the saying goes, misfortunes rarely come alone: in 1998 the country was sent into a catastrophic economic crisis.

In the 1990s, the Global North used the economic crises to export free trade and privatisation to every corner of the world. Defenders of this regime would explain the crises of the 1990s in Latin America, Asia and Russia as being caused by corrupt politicians or crazy speculation. However, this is not entirely convincing. Neither corruption nor speculation first appeared in the 1990s, and they certainly have not disappeared since. These are also explanations that place the blame on the countries affected (corruption) and reduce the crises to psychological and individual phenomena (speculation). Thus, either no one or only the affected countries are at fault.

Anti-inflation rhetoric was at the core of neoliberal policies. One just 'had to' smash unions, cut wages, reduce government spending and restructure financial and economic policy. In the shadow of all the rhetoric and theory, a particular type of inflation was cultivated: the price of assets such as housing and shares. Andrew Sayer calls this the dirty secret of neoliberalism. If food prices had risen as much as house prices in the UK over the past forty years, a chicken in 2015 would cost £51.18.[37] The US Federal Reserve tried to strengthen its economy through – in the words of Robert Brenner – the 'bizarre way' of nurturing asset-price bubbles which Brenner calls 'asset-price Keynesianism'.[38] Now it was not higher wages that would ensure growth, as in classical Keynesianism, but rising share and house prices. Lower interest rates would force asset prices up, and owners would be able to borrow more and more. This first culminated in the high-tech boom that burst with the so-called IT crisis of 2000–01, but the problems were quickly solved by a new and bigger bubble in the housing sector.

On the morning of 20 September 2008 – in the days that followed the collapse of Lehman Brothers, the fourth-largest investment bank in the

36 Magnusson, *Finanskrascher*, p. 200.

37 Andrew Sayer, *Why We Can't Afford the Rich* (Bristol: Policy Press, 2015), pp. 97, 101, 103.

38 Robert Brenner, 'What's Good for Goldman Sachs Is Good for America: The Origins of the Current Crisis', Institute for Social Science Research working paper, UCLA, 2009.

US – the US Treasury secretary Hank Paulson warned US Congress that the world economy was at risk of collapse within twenty-four hours. US Federal Reserve chairman Ben Bernanke said in a meeting with senior members of Congress that, without immediate policy action, 'we may not have an economy by Monday'.[39] The message from Paulson and Bernanke was simple. The entire financial system needed to be saved and at any cost. Once again, realpolitik trumped economic theory. All ideological talk about states not being active players under neoliberalism disappeared when it came to preventing an implosion of the entire economic system.

This time, the crisis was triggered by a bursting US housing bubble. When the crisis broke, the absurdity of a system based on giving mortgages to people who could not afford them was exposed. Banks could do this and avoid the risk, as loans were repackaged and sold. Big rating agencies – Fitch, Moody's and Standard & Poor's – did not want to lose customers, and since mortgages had become a huge source of revenue, they continued to give top ratings to these packages. Of all the housing packages rated triple-A in 2006 – i.e., highest quality and lowest credit risk – no less than 73 per cent was downgraded to junk in 2010.[40]

The crash could be felt worldwide, as the securitised mortgages formed a globally interconnected system. But this was more than a bursting housing bubble. Most commentators would agree it was a *financial* crisis, as a financial system based on banks depending on borrowing/lending from/to each other broke down.[41] But the 2008 crisis was also something more than a financial crisis. It appeared as a financial crisis, for finance was the most dynamic sector, and it was here that the underlying weakness first manifested itself.[42] An economy better at

39 Adam Tooze, *Crashed: How a Decade of Financial Crises Changed the World* (New York: Viking, 2018), p. 162. See also Skidelsky, *Keynes*, pp. 9–10.

40 Financial Crisis Inquiry Commission, *The Financial Crisis Inquiry Report: Final Report of the National Commission on the Causes of the Financial and Economic Crisis in the United States* (Washington, DC: US Government Printing Office, 2011), p. 122. See also, e.g., Manuel B. Aalbers (ed.), *Subprime Cities: The Political Economy of Mortgage Markets* (Malden, MA: Wiley-Blackwell, 2012); David Harvey, *The Enigma of Capital: And the Crises of Capitalism* (London: Profile Books, 2010); McNally, *Global Slump*, chs 4–5.

41 Tooze, *Crashed*, p. 6. See also Perry Anderson, 'Situationism à L'envers?', *New Left Review* 119 (2019), p. 71.

42 Paul Mattick Jr, *Business as Usual: The Economic Crisis and the Failure of Capitalism* (London: Reaktion Books, 2011), p. 25. On how crises are often framed as credit, monetary and financial crises, see also Marx, *Capital*, vol. 3, p. 621; David

creating bubbles than generating profits in productive sectors had seemingly come to an end. The next question then becomes: Why was the underlying economy relatively weak?

Socialists were far better prepared for the 2008 crisis than they were in 1973, at least theoretically. Alex Callinicos, Gérard Duménil and Dominique Lévy, John Bellamy Foster and Fred Magdoff, Guglielmo Carchedi, David Harvey, Andrew Kliman, Paul Mattick Jr, David McNally, and Michael Roberts, among others, all published books on the crisis between 2009 and 2012. A few different explanations were provided; John Bellamy Foster and Fred Magdoff mobilised stagnation theory, Sam Gindin stressed that this was primarily a financial crisis, and Brenner saw *overcapacity* in global manufacturing as a main cause, due to an intensification of capitalist competition.[43] But the two main positions that emerged were overproduction and TRPF.

The most famous proponent of the overproduction thesis was David Harvey. After the crisis of the 1970s onward, it was possible to produce cheaper in Asia (places of production), but workers in the West (places of consumption) did not receive sufficient wage increases to buy the increasing amount of goods. This lack was compensated by loans and increased debt. Excess capital was sloshing around in search of places to be realised, and, as the human geographer he is, Harvey stresses how this leads to overinvestment in urban construction; and so it was no coincidence that the crisis began precisely in the housing sector.[44]

With the theory of the falling rate of profit, there was little doubt that the organic composition had increased sharply, as the number of workers per million dollars of assets in the US productive sectors declined

McNally, 'Beyond the False Infinity of Capital: Dialectics and Self-Mediation in Marx's Theory of Freedom', in Albritton Robert and John Simoulidis (eds), *New Dialectics and Political Economy* (London: Palgrave, 2003), pp. 12–13.

43 John Bellamy Foster and Fred Magdoff, *The Great Financial Crisis: Causes and Consequences* (New York: Monthly Review Press, 2009); Sam Gindin, 'Clarifying the Crisis', *Jacobin* (2014); Brenner, 'What's Good for Goldman Sachs'.

44 Harvey, *The Enigma*; David Harvey, 'History versus Theory: A Commentary on Marx's Method in Capital', *Historical Materialism* 20, no. 2 (2012), pp. 3–38. It is worth noting that Harvey often refers to overaccumulation rather than overproduction. When we talk about this as a general phenomenon, we will in the following use the established term 'overproduction'.

between 1947 and 2010 from sixty-five to six.[45] Although the profitability of many financial firms was huge until 2008, the rate of profit in productive sectors and firms was less impressive. And with a lower rate of profit came higher instability.

After 2008, there were discussions on how to measure the rate of profit, how to periodise the ups and downs, how to read statistics, and what the direct or indirect causes of the crisis were.[46] For example, if lower wages are a countervailing trend, could the TRPF have been an active force during neoliberalism? Here we must remember that trends and counter-trends are not mutually exclusive. It is not the case that either a tendency acts or its counter-tendency acts. While capitalists were rationalising and implementing new technologies (leading to higher organic composition), the counter-tendencies to TRPF were also very strong – such as more intensive exploitation of labour, speculation in unproductive sectors (especially real estate and finance capital) and often lower interest rates.[47] Such mechanisms kept neoliberalism alive. Yet, instability could not be solved, only postponed.

Unfortunately, the debates between advocates of overproduction and the TRPF have been very polarising. The truth is that both tendencies have something important to tell us about the economic crisis of capitalism. Rather than mutually exclusive theories – as if this were the Champions League of Marxist crisis theory – overproduction and TRPF should be seen as two vantage points from which we can analyse the world.

45 Guglielmo Carchedi and Michael Roberts, 'A Critique of Michael Heinrich, Crisis Theory, the Law of the Tendency of the Rate of Profit to Fall, and Marx's Studies in the 1870s', *Monthly Review*, 1 December 2013; Guglielmo Carchedi and Michael Roberts, 'The Long Roots of the Present Crisis: Keynesians, Austerians, and Marx's Law', *World Review of Political Economy* 4, no. 1 (Spring 2013).

46 On periodisation see, e.g., David McNally versus Joseph Choonara in *International Socialism*, nos 134 and 135. On the role of TRPF, see also Deepankar Basu and Ramaa Vasudevan, 'Technology, Distribution and the Rate of Profit in the US Economy: Understanding the Current Crisis', *Cambridge Journal of Economics* 37 (2013), pp. 57–89; contra Michael Roberts, 'The Rate of Profit (Again!)', *The Next Recession*, 31 August 2011. See also Carchedi and Roberts ('A Critique of Michael Heinrich') contra Michael Heinrich, 'Crisis Theory, the Law of the Tendency of the Profit Rate to Fall, and Marx's Studies in the 1870s', *Monthly Review* 64, no. 11 (April 2013), pp. 15–31.

47 See also Carchedi and Roberts, 'A Critique of Michael Heinrich'; Carchedi and Roberts, 'The Long Roots'; Kliman, *The Failure of Capitalist Production*; Shaikh, *Capitalism*.

Monocausality – and the Critique (of the Critique)

One reason why Marxists have quarrelled over which theory of crisis is the right one is that Marx did write a few different things himself. In the third volume of *Capital*, Marx writes that the 'ultimate reason for all real crisis always remains the poverty and restricted consumption of the masses', which points to the theory of underconsumption. In the *Grundrisse*, it is the tendency of the rate of profit to fall that is the focus: 'in every respect the most important law of modern political economy'. But here, we also read that production will not expand simultaneously and in equal proportions in all branches and spheres of production, which points to the theory of disproportionality – as does Marx's argument in *Theories of Surplus-Value* that overinvesting in one sector and underinvesting in others can create particular crises. While overproduction, Marx writes in *Theories of Surplus-Value*, is 'the basic phenomenon in crises'.[48]

Since much of Marx's writings were published after his death and often edited by others, the debates are often about what Marx *really* meant – or did not mean. One criticism of Marx's texts on overproduction is that they were mostly polemics against Sismondi and Malthus, while one criticism of TRPF was that Marx's texts on this are really just reflections and notes on the theories of Ricardo.[49]

In addition to passages emphasising the importance of individual crises, we also find a different approach in Marx's *Theories of Surplus-Value*. Here, we read that in major crises – something Marx calls the 'world market crises' – 'all the contradictions of bourgeois production erupt collectively'. This is in contrast to '"particular crisis" (*particular* in their content and in extent) [where] the eruptions are only sporadical, isolated and one-sided'.[50] Rather than searching for one theory, the focus now becomes how different theories and contradictions intertwine.

48 Marx, *Capital*, vol. 3, p. 615; Marx, *Grundrisse*, pp. 414, 748; see also Clarke, *Marx's Theory*, p. 142; Karl Marx, *Theories of Surplus-Value*, part 2 (Moscow: Progress Publishers, 1968), pp. 521, 528. In Marx's grand and original plan for *Capital*, the sixth and final part was to be about the world market and crises, cf. Sven-Eric Liedman, *A World to Win: The Life and Works of Karl Marx* (London: Verso, 2018), pp. 370–3.

49 See Makoto Itoh, cited in Bensaïd, 'The Time of Crises', p. 29; see David Harvey, 'Debating Marx's Crisis Theory and the Falling Rate of Profit, Draft', in Turan Subasat and John Weeks (eds), *The Great Meltdown of 2008: Systemic, Conjunctural or Policy Created?* (Northampton, MA: Edward Elgar, 2014).

50 Marx, *Theories of Surplus-Value*, p. 534.

The Belgian Marxist Ernest Mandel is perhaps the most famous critic of monocausality, that is, the view that one tendency governs the economic crises of capitalism.[51] Inspired by Mandel, we will work here on the hypothesis that overproduction and TRPF are two starting points for understanding crises. We can call them realisation and production perspectives. From the realisation perspective, we see how the contradiction between production and realisation creates tendencies towards overproduction. From the production perspective, we see how new 'improvements' in production tend to increase productivity and the organic composition of capital. The 'practical capitalist', however, Sweezy reminds us, is unlikely to see any difference: 'For him the trouble is always insufficient profitability from whatever source it may arise.'[52] But when analysing crises, the realization and production perspectives are distinct viewpoints that highlight different phenomena and provide us with varying insights.

I think the *best* starting point is to try to understand that/how crises are constituted by different contradictions. But perhaps we should rather say that this starting point is the *least bad*. Even this position has its challenges. The hardest thing is to determine how many and what tendencies are at play. Mandel was adamant that there are various tendencies that need to be explained, but unsure of exactly which ones. Jan Willem Stutje argues that Mandel operated with five tendencies in the *Frankfurt Colloquium*, six in *Late Capitalism*, five in *Long Waves of Capitalist Development*, and in 1984 he operated with ten.[53]

Another challenging question is how different trends are related. A quick answer is that everything is potentially relevant and must be understood dialectically, and that nothing completely determines

51 Mandel, 'Introduction', in Marx, *Capital*, vol. 3, pp. 38–53; Mandel, *Late Capitalism*, pp. 34–43. For defences of monocausality, see Guglielmo Carchedi, 'Zombie Capitalism and the Origin of Crises', *International Socialist Journal* 125 (2010); Carchedi, *Behind the Crisis*. For discussions, see also McNally, *Beyond the False Infinity*, p. 7; McNally, *Global Slump*, p. 29; Chris Freeman and Francisco Louçã, *As Time Goes By: From the Industrial Revolutions to the Information Revolution* (Oxford: Oxford University Press, 2001), pp. 80, 91; Bob Jessop, 'The Symptomatology of Crises, Reading Crises and Learning from Them: Some Critical Realist Reflections', *Journal of Critical Realism* 14, no. 3 (2015), p. 250; Clarke, *Marx's Theory*, ch. 9.

52 Sweezy, *The Theory of Capitalist Development*, p. 146.

53 Jan Willem Stutje, *Ernest Mandel: A Rebel's Dream Deferred* (London: Verso, 2009), p. 133.

everything else. But simple answers often provide only short-term satisfaction. Even if everything is related, some things are so irrelevant that they should not be included in the analysis. To seek deeper answers, we must proceed with two different methodologies: one is more theoretical and analyses the phenomenon at different levels of abstraction, the other is more empirical and historical. Only through such analyses can we approach the way different components interact dialectically and create general profit rates, levels of growth – and not least crises.

Another difficult issue is the temporality of crises. History shows that major economic crises do not come every year, but they do recur. In the nineteenth century, crises seemed to come in cycles of about seven to ten years, which Marx related to the lifespan of machines. Today, such cycles are discussed especially in the construction sector.[54] Here, we must make a distinction between (often minor) crises within a hegemonic order and (often major) crises that challenge the prevailing organisation of capitalism.[55] In the former, prevailing class relations and institutional frameworks may largely remain unchanged. Economic cycles in construction are more or less ingrained throughout the industry. Black Monday on 19 October 1987 was one of the largest stock market crashes capitalism has ever seen, but passed relatively quickly. The crises in Asia in the 1990s were very large in scale and impact, but could be used by the ruling class to legitimise the prevailing hegemony.

Crises that challenge the hegemonic order reveal what Mandel called 'qualitative leaps' in the organisation of work and technology.[56] Defining years are 1873, 1929, 1973 and 2008, and these crises developed into prolonged depressions/recessions, or at least periods when it was difficult to re-establish growth and rate of profit. They also included several distinct crises. The problems of 1873 lasted until 1897; the Great Depression ran from 1929 to about 1939; the Great Stagflation from 1973 to 1981; and the political-economic problems that surfaced in

54 Mandel, *Late Capitalism*, ch. 4. For a discussion of the construction sector, see Jamie Gough, *Work, Locality and the Rhythms of Capital: The Labour Process Reconsidered* (London: Psychology Press, 2003).

55 See also Elmar Altvater, *The Future of the Market: An Essay on the Regulation of Money and Nature after the Collapse of 'Actually Existing Socialism'* (London: Verso, 1993), pp. 42–9.

56 Mandel, *Late Capitalism*, p. 112.

2008 are still with us. The longer periods of crisis also seem to consist of two shorter periods of deeper crises: 1929–33 and 1937–39; 1973–74 and 1979–81; 2007–10 and 2020–?[57]

Deeper crises are more like what Gramscians call organic crises (when several crises coincide and merge) and hegemonic crises (when prevailing ways of organising production, consumption and circulation are challenged and need to be changed). How long do such crises last? According to Elmar Altvater, a crisis lasts until the restructuring process is complete; he points to Antonio Gramsci's famous dictum: 'that the old is dying and the new cannot be born'.[58]

On the one hand, economic crises are very much shaped by people and politics, so there is no basis for believing in a determinate temporality, that crises recur exactly every *x* year with mechanical precision. On the other hand, since economic crises very much arise from contradictions within capitalism, it would be stupid not to compare current with past temporality. To explain the intervals between major crises, which occur roughly every forty to sixty years, the term 'long waves' is often used. I will admit to a touch of agnosticism regarding wave theory. Arguing in favour of 'long waves', we may add that, empirically, there seems to be some kind of pattern, which may stem from general rates of profit and hegemonic systems within capitalism. Arguing against 'long waves', we have the fact that it is very possible to make different waves. Depending on whether we take profit rate, technology, politics or whatever as our starting point, we can use statistics and history to create quite different waves.[59]

Theories of long waves would be easier to understand if there were *one* cause, *one* driver and *one* solution to the crises. Not the *multiple* tendencies that is our hypothesis here. If we are to use the term 'long waves', it must be as a metaphor for the development of capitalism, rather than a model for accurately predicting crises over the next hundred years. Development must be understood non-mechanically,

57 McNally, *Global Slump*, p. 39; see also Roberts, *The Long Depression*, p. 229.

58 Altvater, *The Future of the Market*; Antonio Gramsci, *Selections from the Prison Notebooks* (London: Lawrence and Wishart, 1971), p. 276.

59 See, e.g., Mandel, *Late Capitalism*, p. 247; Carlota Perez, *Technological Revolutions and Financial Capital: The Dynamics of Bubbles and Golden Ages* (Cheltenham: Edward Elgar Publishing, 2002). For my somewhat ambivalent relationship to long wave theory, see Holgersen, *Staden*, ch. 2.

less as 'true cycles' and more as 'historical epochs'.[60] This is also how I read Mandel's description of waves. The question is then, of course, whether 'waves' is the best term.

Capitalism certainly comes with calmer and more turbulent periods. A more fruitful way of capturing the (un)regular rhythm of capitalism is with the concept of hegemonic periods. Different economic trends create a common profit rate and are important for driving technological innovations, while when, where and how crises occur also depends on class struggle and on the historical-geographical organisation of capitalism. That these periods are not completely determined by objective and soulless economic tendencies becomes especially clear when we discuss how economic crises are resolved. The question of why there is a crisis is only half the story. I think Bill Dunn is correct when pointing out that most Marxists focus more on why crises occur than on how crises are resolved.[61] If we want to understand how capitalism survives because of its crises, we must turn to the equally important question of how economic crises are resolved.

Solving a Crisis/Crisis as Solution

When we focus only on how economic crises are created, capitalism appears brittle, fragile and weak, but, when we emphasise how they are resolved, it is difficult not to be overwhelmed by how flexible, viable and adaptable capitalism actually is. The ruling class under capitalism, for example, seems able to accept everything from a welfare state, human rights and better working conditions to poverty, genocide and fascism in order to maintain its power.

60 Mandel, *Late Capitalism*, pp. 112, 140. We may add a 'fun fact' – for lack of a better word – about wave theory: Stalinism is not exactly known for fostering critical thinking, and Stalin was no great believer in the notion that the economy operated in cycles, preferring rather the theory that capitalism would fall. This became part of the explanation for why Soviet economist Nikolai Kondratieff – the father of wave theory – was assassinated by Stalin's henchmen in 1938. See Paul Mason, *Postcapitalism: A Guide to Our Future* (London: Allen Lane, 2015), pp. 32–4; Kjell Östberg, 'Den solidariska välfärdsstaten och förändringarna i den politiska dagordningen', in Torsten Kjellgren (ed.), *När skiftet äger rum: vad händer när den politiska dagordningen ändras* (Stockholm: Tankesmedjan Tiden, 2017), p. 26.

61 Dunn, 'Marxist Crisis Theory', p. 524.

Economic crises are more than just shock and expressions of underlying contradictions: under capitalism the crises are also *solutions* to the crises. There are underlying tendencies and contradictions in the political economy itself that gradually make the economy more unhealthy, and this tends to escalate to a point: the crisis. It is through crises that stagnant and crisis-prone economies find new opportunities for profit and growth.

In the third volume of *Capital*, we read that crises are 'never more than momentary, violent solutions for the existing contradictions, violent eruptions that re-establish the disturbed balance for the time being.'[62] Here Marx is basically spot on. Capitalism needs economic crises to act as a valve to relieve pressure when underlying contradictions are boiling, and to assist the reshaping and restructuring of the system when existing hegemonic orders have come to an end. Crises become capitalism's own problem solvers. This may sound counterintuitive, but in context of the capitalist system, they are also the starting point for new investments and new periods of increased growth. The concept of creative destruction can help us understand how this happens.

Creative Destruction

The German economist Werner Sombart wrote in 1913 about a 'creative spirit in destruction', and Joseph Schumpeter popularised the concept thirty years later.[63] Since then, the idea of creative destruction has been used by people with different theoretical and political orientations. The popularity of the term may be due to its catchiness, but it also points to something very central: capitalism renews itself through economic crises.

Creative destruction will here be understood as a process constituted by three interlinked moments. The first is the devaluation and destruction of capital; the second is the process by which crises are solved using new ways of organising the economy, new class relations, new technologies and new regulations replacing the old ones; and the third is new

62 Marx, *Capital*, vol. 3, p. 357; see also Mandel, *Late Capitalism*, p. 104; Harvey, *The Enigma*; Štajner, *Crisis*, pp. 49–50; Marx, *Capital*, vol. 2, p. 264.

63 Perez, *Technological Revolutions*, p. 22; Joseph Schumpeter, *Capitalism, Socialism, and Democracy* (New York: Harper & Brothers, 1942), ch. 7.

investments. Creative destruction is thus an intertwining of disruptive, transformative and constructive processes. To simplify, we can add that Marxists emphasise the first, liberals inspired by Schumpeter love the second, and Keynesians worship the third. But, in fact, these are not three completely independent processes. One part of the process cannot happen without the others, and it may often be unclear what is destructive and what is creative.

The first and most important part is the destruction. It can occur in various ways, such as companies going bankrupt, the value of money being eroded (through inflation or default on debts), goods being sold at below-cost or remaining unsold and being discarded, or productive capacity being idle, abandoned or destroyed. Politicians can also choose to devalue the currency (known as external devaluation) or by cutting and tightening, for example, people's wages, pensions and welfare (also known as internal devaluation). An extreme variant is destruction of capital through war. It is in such situations, says Marx, that both use-value and exchange-value 'go to the devil'.[64] When enough capital has been destroyed, profit rates can start at a higher level and the problems of overproduction are restored. It matters both *how much* and *what kind* of capital is destroyed.[65]

The second part of creative destruction is transformational. For minor cyclical fluctuations, government stimulus policies may do the work, but for major transformations – like in the 1930s, 1970s or 2020s – something far more is needed. Then you cannot escape new rounds of restructuring. This is when yesterday's power relations are questioned, new ideologies, geographies and technologies replace old ones and the political economy is transformed. In these new landscapes, capitalists and even states that have survived the crisis (relatively unscathed) can buy up old companies, banks, factories, land, etcetera, and employ people to make profits again, in new ways.

The third part of creative destruction is new investments. Where increased private investments can be a sign that the crises are coming to an end, large public investments can be part of the very solution to the

64 Marx, *Theories of Surplus-Value*, p. 496. On devaluation and crisis, see, e.g., Mattick Sr, *Marx and Keynes*; Shaikh, 'An Introduction'; Harvey, *The Limits*; Clarke, *Marx's Theory*; Roberts, *The Great Recession*; Kliman, *The Failure of Capitalist Production*.

65 See, e.g., Kliman, *The Failure of Capitalist Production*; Mandel, *The Second Slump*, pp. 78, 80.

crisis. For a Keynesian like Paul Krugman, the conclusion is simple: extensive government counter-cyclical policies can revive demand and growth, and if they have not succeeded, the government should borrow 'until the private sector is ready to carry the economy forward again.'[66] And state interventions *are* crucial in crisis, but massive government investment *alone* does not solve deep crises. There is always a risk of throwing money at the wrong sectors. State investment can indeed push problems forward rather than contributing to new rounds of profits and growth. Soon, you might be back to similar problems, but with bigger government deficits and more public debt.[67]

Creative destruction is the disruption of old economic, social and cultural orders. Major economic crises reorganise, restructure and reshape; they modernise the political economy and move on. Crises are not only problems to be solved, but also solutions to be used. That a large number of people might be consigned to the historical dustbin, to paraphrase Stuart Hall, is obvious. This is the 'law' of capitalist modernisation, according to Hall: 'uneven development, organised disorganisation.'[68]

We will further scrutinise this and much more in the following, but, before we go any further with economic crises, we also need to introduce the ecological ones.

66 Krugman, *End This Depression*, p. xi.

67 On how this happened in Sweden in the 1970s, see Holgersen, *Staden*, chs 3 and 5.

68 Stuart Hall, 'Gramsci and Us', in *The Hard Road to Renewal: Thatcherism and the Crisis of the Left* (1988), published on versobooks.com, 10 February 2017.

2
An Infected Wound

It is fascinating – or grotesque – to see the amount of money and wealth that capitalism has created for some. It is even more grotesque – or fascinating – to see how extreme poverty and ecological crises are produced simultaneously. In the age of contradictions, advances in human rights come at the same time as new genocides. Every second, some fifty tonnes of food are thrown away or destroyed, about one-third of all food produced in the world, while 821 million people are malnourished. In cities, people sleep in the streets outside empty houses. Shopping malls sell expensive products that employees inside said malls cannot afford to buy. On the Hong Kong–Macau Expressway, fifty cars can drive next to each other, in the same direction, yet there are traffic jams.[1]

In a speech in London in 1856, Marx said that in our days, everything seems pregnant with its contrary:

> The victories of art seem bought by the loss of character. At the same pace that mankind masters nature, man seems to become enslaved to other men or to his own infamy. Even the pure light of science seems unable to shine but on the dark background of ignorance.[2]

1 Magnus Nilsson, 'En tredjedel av all mat slängs', sverigesnatur.org, 3 October 2018; Jim Gorzelany, 'The World's Worst Traffic Jams, Ever', *Forbes*, 15 October 2015.

2 Quoted in Marshall Berman, *All That Is Solid Melts into Air* (London: Verso, 1983), p. 20.

Marx captured many of the contradictions of capitalism, even between nature and capitalism: that capitalist agriculture could produce more and more, but only at the price of 'undermining the original sources of all wealth: the soil and the worker'.[3] But Marx could hardly have imagined the depth of this contradiction, namely that the progressive forces of capital would change the climate.

During the Holocene geological epoch, which began 11,700 years ago, temperatures have varied by a maximum of one degree above or below the average. This means that humans developed agricultural societies, cities and states and what we call civilisations over a period of time that was, climatically, relatively stable. The regulating capacity of the earth enabled so-called human development.[4] This does not mean that past rulers were more concerned about the environment; they simply lacked the technological preconditions and the political-economic system – a growth-dependent system, greased with coal and oil – that the ruling class under capitalism has had the luxury of controlling. Environmental changes existed, but most were 'natural' and relatively small, and even when human activity was a contributing factor to changes, the consequences were limited. Nothing that can be compared to what has happened under capitalism.

Now, all stability is lost in just a few generations. Between 1906 and 2005, temperatures rose by almost one degree. And this is only the beginning; even if we start gradually reducing greenhouse gas emissions today – which we will not – projections show that, by 2070, it will be warmer than at most, if not all, times since modern humans emerged as a species 200,000 years ago.[5] Now, for the first time, a biological species (humans) will be responsible for the mass extinction of other species. Species are dying a thousand times faster than they did before capitalism, and the killing is happening far faster today than the previous mass

3 Karl Marx, *Capital*, vol. 1 (London: Penguin, 1976), p. 638.

4 Johan Rockström et al., 'A Safe Operating Space for Humanity', *Nature* 461 (2009), p. 472; Andreas Malm, *Corona, Climate, Chronic Emergency: War Communism in the Twenty-First Century* (London: Verso, 2020), p. 117. There were much greater variations before the Holocene. During the Pleistocene – from about 2,580,000 years ago to about 11,700 years ago – temperatures varied by five to ten times more than during the Holocene. As long as temperatures were relatively unstable, we were hunters and gatherers.

5 Colin Waters et al., 'The Anthropocene Is Functionally and Stratigraphically Distinct from the Holocene', *Science* 351, no. 6269 (2016).

extinction 65 million years ago, when a giant meteorite struck the Yucatan Peninsula and wiped out 75 per cent of all species on Earth. Including the dinosaurs.[6]

If feudalism had never been replaced by capitalism, but people nonetheless dug up more and more fossil fuels, we could talk about the *crises of feudalism*. That did not happen. If the Soviet Union had won the Cold War, perhaps we could have discussed climate change in terms of a *crisis of Soviet dictatorship*. That did not happen either. It is perfectly legitimate to object to my focus on capitalism and argue that environmental destruction in the Soviet Union was also extensive. Other social systems can also create ecological problems. But feudalism is dead and the Soviet Union is long gone. The political economy we have to deal with today is capitalism, whether we like it or not. Climate change began under capitalism, it is developing and escalating under capitalism, and its solution is being attempted – but monumentally failing – within capitalism.

Ecological Crisis

'Labour is the source of all wealth and culture', proclaimed the so-called Gotha Programme – the German Social Democratic Party programme adopted in Gotha in 1875. Marx objected: 'Labor is *not the source* of all wealth. Nature is just as much the source of use-values (and it is surely of such that material wealth consists!) as labor, which itself is only the manifestation of a force of nature, human labor power.'[7] The fact that nature does not by itself create value – from the perspective of capitalism – does not mean that it is not absolutely central to our political economy, and so clearly *valuable* from a human perspective.

In the young Soviet state, advanced debates were held, and with Lenin's support, nature reserves were set up in the midst of a burning

6 Sverker Sörlin, *Antropocen: en essä om människans tidsålder* (Stockholm: Weyler förlag, 2017), pp. 147, 152; Ian Angus, *Facing the Anthropocene: Fossil Capitalism and the Crisis of the Earth System* (New York: Monthly Review Press, 2016), p. 73.

7 Karl Marx, *Critique of the Gotha Programme*, in *Karl Marx and Frederick Engels: Selected Works*, vol. 3 (Moscow: Progress Publishers, 1970 [1875]), pp. 13–30. See also David McNally, 'Beyond the False Infinity of Capital: Dialectics and Self-Mediation in Marx's Theory of Freedom', in Albritton Robert and John Simoulidis (eds), *New Dialectics and Political Economy* (London: Palgrave, 2003), p. 13.

civil war.[8] But it was technological and developmental optimism – or shall we say fetishism? – rather than environmentalism that became the main track in the Soviet Union. Socialism would be *built* – even with airplanes, space rockets and nuclear power plants. We cannot escape the fact that, for much of the twentieth century, Marxism was associated with regimes that were hardly stars in any ecological sky.

When Marxists began to discuss ecology again in the 1960s, it was on the basis that Marxism was as much part of the problem as the solution. The best-known figure in what is often called the first wave of eco-Marxists was the North American sociologist James O'Connor, and his theory of capitalism's second main contradiction was groundbreaking.[9] Capitalism's first contradiction was that between the productive forces and the relations of production – that is, the tension between the concentration of profits and power on the one hand, and the increasing socialisation of production on the other (which, for O'Connor, leads to crises of overproduction). The second contradiction was between the productive forces and the *conditions* of production. (O'Connor built on Marx and drew on Karl Polanyi when arguing that three such conditions of production were external physical conditions, labour power and the common, general conditions of social production.[10]) As is so typical of first-wave eco-Marxists, O'Connor tried to elaborate on what he thought Marx had missed: Marx was not wrong, he was only half right. Where Marx and Marxists had focused on labour versus capital, the focus on nature now opened up a parallel path to socialism.[11]

8 On ecology in early Marxists, see Arran Gare, 'Soviet Environmentalism: The Path Not Taken', *Capitalism Nature Socialism* 4, no. 3 (1993), pp. 69–88; Angus, *Facing the Anthropocene*, pp. 108–11; Rikard Warlenius, 'Inledning: Fyra debatter och en begravning', in Rikard Warlenius (ed.), *Ecomarxism: Grundtexter* (Stockholm: Tankekraft, 2014), pp. 13–14; John Bellamy Foster, Brett Clark and Richard York, *The Ecological Rift: Capitalism's War on the Earth* (New York: Monthly Review Press, 2010), ch. 11; Kate Soper, 'En grönare Prometheus: Marxism och ekologi', in Warlenius, *Ecomarxism*, pp. 177–82.

9 James O'Connor, 'Capitalism, Nature, Socialism: A Theoretical Introduction', *Capitalism Nature Socialism* 1, no. 1 (1988), pp. 15–16; see also James O'Connor, 'On the Two Contradictions of Capitalism', *Capitalism Nature Socialism* 2, no. 3 (1991), pp. 107–9.

10 See O'Connor, 'Capitalism, Nature', p. 13; see also Karl Polanyi, *The Great Transformation: The Political and Economic Origins of Our Time* (Boston: Beacon Press, 2001 [1944]); Alan P. Rudy, 'On Misunderstanding the Second Contradiction Thesis', *Capitalism Nature Socialism* 30, no. 4 (2019), pp. 20–2.

11 O'Connor, 'Capitalism, Nature', pp. 13–14, 17, 31.

This perspective was criticised by the next generation of ecological Marxists who emerged in the second half of the 1990s. They argued that such a focus on two contradictions risked ignoring how these constantly interacted, and more succinctly, how the exploitation of labour and the destruction of nature were embedded in each other. Rather than lashing out at Marx and Marxism for lacking an ecological focus, second-generation Marxists were fostering new interpretations of Marx and Engels, and original texts were used to directly understand ecological crises.

There is no intrinsic value in making Marx more or less of an eco-socialist than he really was. We must accept there are different ways of reading Marx; different notions of history and socialism coexist in embryonic form in Marx's works.[12] It is of less importance whether Marx was *really* an eco-socialist, a degrowther, left-productivist or a Promethean. What is crucial, however, is that many of Marx's analyses of capitalism are absolutely central if we truly want to understand what is happening.

New Rifts and New Fuel

Humans and nature have had a reciprocal relationship for as long as humans have existed. Humans must eat, make tools and build homes from what nature can provide, and excrement maintains the fertility of the earth. *For dust thou art, and into dust thou shalt return.*

In the nineteenth century, the German agricultural chemist Justus von Liebig explained the interaction and biochemical exchange between plants, animals and soil with the then new word 'metabolism' (*Stoffwechsel*). Marx took Liebig's work from an individual to a societal scale, and a specific way of regulating the general process that binds man and nature mutually to each other is thus called *social metabolism*.[13]

12 Enzo Traverso, *Understanding the Nazi Genocide: Marxism after Auschwitz* (London: Pluto Press, 1999), p. 23.

13 For discussions, see John Bellamy Foster, *Marx's Ecology: Materialism and Nature* (New York: NYU Press, 2000), ch. 5; Foster, Clark and York, *The Ecological Rift*, part 1; Warlenius, 'Inledning', pp. 31–3; Angus, *Facing the Anthropocene*, ch. 7; Andreas Malm, *The Progress of This Storm: Nature and Society in a Warming World* (London: Verso, 2018), ch. 6; Kohei Saito, *Marx in the Anthropocene: Towards the Idea of Degrowth Communism* (Cambridge: Cambridge University Press, 2022). For discussions in Marx, see, e.g., Marx, *Capital*, vol. 1, ch. 15; Karl Marx, *Capital*, vol. 3 (London: Penguin Classics, 1981), ch. 47. For a critique of the theory as it separates 'man' from 'nature', see

Different historical systems organise this in different ways. With capitalism – with its perpetual pursuit of profit and compulsion to constant accumulation – there arises for the first time in human history what we might call a general and large-scale *rift* or *crack* in the social metabolism between humans and nature.

One of the most worrying environmental issues of the nineteenth century in both Europe and North America was the diminishing fertility of the soil in the emerging capitalist agriculture. The solution was to import guano from South America and later the introduction of synthetic fertilisers. Modern capitalist agriculture could emerge, producing food for profit, which could further escalate with increased mechanisation, concentration of animals in massive farms, genetic modification of animals, monocultures and increasingly intensive use of chemical inputs. The contradiction was noted by Marx: 'All progress in increasing the fertility of the soil for a given time, is a progress towards ruining the lasting sources of that fertility.' With people being forced from the countryside to the cities, this created a double problem: human excrement that no longer contributed to fertilising the soil for agriculture now created waste and pollution in cities. Marx noticed that 'they can do no better with the excrement produced by 4.5 million people than to pollute the Thames with it, and at enormous cost.' Engels remarked that more manure was dumped into the river every day than the entire kingdom of Saxony could produce.[14]

The rift between humans and nature has manifested itself in different ways. What scientists call planetary boundaries can be seen as concrete expressions of metabolic rifts between society and nature.[15]

Jason W. Moore, 'Toward a Singular Metabolism: Epistemic Rifts and Environment-Making in the Capitalist World-Ecology', *New Geographies* 6 (2014), pp. 10–19; Jason W. Moore, *Capitalism in the Web of Life: Ecology and the Accumulation of Capital* (London: Verso, 2015). On the theory's construction of a 'nostalgia' over pre-capitalist 'natural' metabolic relations, see Matt Huber, *Lifeblood: Oil, Freedom, and the Forces of Capital* (Minneapolis: University of Minnesota Press, 2013), pp. 179–80.

14 Marx, *Capital*, vol. 3, p. 195; Friedrich Engels, *The Housing Question* (London: Martin Lawrence, 1942), p. 95. See also Marx, *Capital*, vol. 1, p. 638; Foster, *Marx's Ecology*, pp. 149, 163.

15 See John Bellamy Foster, 'The Epochal Crisis', *Monthly Review* 65, no. 5 (2013), p. 1. Rockström and colleagues identify nine boundaries: 1) climate change; 2) rate of biodiversity loss (terrestrial and marine); 3) interference with the nitrogen and phosphorus cycles; 4) stratospheric ozone depletion; 5) ocean acidification; 6) global freshwater use; 7) change in land use; 8) chemical pollution; and 9) atmospheric aerosol

Overfishing is by definition taking more from nature than it (the fish) can reproduce. Nuclear power is a fascinating phenomenon. While waste, historically, has been something that nourishes the earth, radioactive waste must be stored in isolation for hundreds of thousands of years before it is harmless to humans and nature. Capitalist agriculture does not primarily produce food, it produces profit. To paraphrase Richard Lewontin: while past agriculture was about growing peanuts, industrial agriculture is about converting petroleum into peanuts.[16] Nature's own metabolism of carbon dioxide in the atmosphere, which had taken place for millions of years, was disrupted when fossil fuels were introduced into the political economy. The metabolic rift became a *carbon rift*, a hell gap.[17]

Capitalism existed before the heyday of coal, but it exploded with coal in the engine. Coal had been used even earlier in history, in medieval London to heat houses, or in the Song dynasty (960–1279) for China's growing iron industry. These remained regional exceptions. Coal was not yet *capital*.[18] Its use was not an essential component of anything resembling capitalism – that is, an ever-expanding system of production that was about to put the whole earth under its feet. Once fossil fuels entered the picture, there was no going back. We know the story. Fossil capital revolutionised the world. Fossil fuels became a central component of the capitalist mode of production and the lifeblood of our modern societies. Oil became a crucial factor for warfare, our society and the economy. It became central to people's everyday lives and is also linked to identity and feelings of freedom, exemplified notably through car culture.[19]

---

loading. Scientists believe that we have already exceeded numbers 1, 2, 5 and 6. See Rockström et al., 'A Safe Operating Space'.

16 Quoted in Angus, *Facing the Anthropocene*, p. 158.

17 Foster, Clark and York, *Ecological Rift*, p. 124; Foster, *Marx's Ecology*, p. 149.

18 Andreas Malm, *Fossil Capital: The Rise of Steam Power and the Roots of Global Warming* (London: Verso, 2016), pp. 51–3, 353; Kohei Saito, 'Marx's Theory of Metabolism in the Age of Global Ecological Crisis, Deutscher Prize Memorial Lecture', *Historical Materialism* 28, no. 2 (2020), p. 15; Will Steffen et al., 'The Anthropocene: Are Humans Now Overwhelming the Great Forces of Nature?', *AMBIO: A Journal of the Human Environment* 36, no. 8 (2007), p. 615.

19 Huber, *Lifeblood*; in relation to oil, identity and gender, see Cara Daggett, 'Petro-masculinity: Fossil Fuels and Authoritarian Desire', *Millennium: Journal of International Studies* 47, no. 1 (2018), pp. 31–4. And in relation to ethnicity/racism, see Andreas Malm and the Zetkin Collective, *White Skin, Black Fuel: On the Danger of Fossil Fascism*

The so-called economic golden age of capitalism was a disaster for the earth. Growth was high, but it ran on fossil fuels. The post-war period is what climate scientists call the beginning of the *Great Acceleration*.[20] Human destruction of nature accelerated on the back of post-war optimism about the future, technological revolutions and increased productivity. Mass production and mass consumption simultaneously promoted economic growth and ecological destruction. The dialectic between productive and destructive was in full play.

The Big Climate and the Little Virus

The climate crisis is far from the only ecological crisis. Ideological defenders of the ruling order often point to another ecological crisis when arguing that we can stop global warming within capitalism. In the post-war period, the use of chlorofluorocarbons (CFCs) became popular in refrigerators, spray cans and fire extinguishers, among other things. Then CFC gases (Freon) were found to deplete the ozone layer. An international agreement signed in 1987 (the Montreal Protocol) saw all countries agree to phase out their use, and the story was over, the problem was solved. Could we not simply sign a similar agreement on fossil fuels? Unfortunately, there are huge differences between these ecological crises. A global ban on CFCs was possible because there were only a handful of companies producing just that, profits were already low and there were existing alternatives that could generate profits.[21] Replacing coal, oil and gas with renewable energy sources is a completely different matter. We have built a global infrastructure on fossil fuels which means that we need to fundamentally transform our societies to get out of the problem.

Another ecological crisis exploded in 2020. Covid-19 was often discussed as some exotic Chinese phenomenon, but rarely as what it was, namely a child of capitalism. Viruses are, of course, much older than capitalism, but evolutionary biologist Rob Wallace and historian Mike Davis have long explained how expanding capitalist production is

---

(London: Verso, 2021). For the sake of transparency, it should be mentioned I am a member of the Zetkin Collective.

20 Steffen et al., 'The Anthropocene'; Angus, *Facing the Anthropocene*, ch. 2; Simon Lewis and Mark A. Maslin, 'Defining the Anthropocene', *Nature* 519 (2015), pp. 176–7.

21 Angus, *Facing the Anthropocene*, pp. 87–8.

driving more and new pandemics. They are coming more frequently: SARS (2002), bird flu (2006), swine flu (2009) and Covid-19 (2019). Almost all cases of new pathogens, Rob Wallace argues, are preceded by a changing economic geography; many emerge precisely at the frontiers of expanding capitalist production.[22]

Capitalist animal husbandry acts as a bridge between pathogens in the wild and humans. When animals are kept close together as livestock in monocultures with low genetic diversity, it is easy for viruses to find their next victim. Animals' immune systems are compromised due to high levels of stress, the rapid turnover of animals means that the most robust pathogens prevail in the selection, and overuse of antibiotics in the livestock industry creates resistant bacteria.[23] Avian influenza (H5N1) is strongly linked to the poultry industry, and swine influenza (H1N1) to the pig industry. With Covid-19, it was not the livestock industry that directly caused the pandemic. Rather, the trail leads to the wild food industry.[24]

Land grabbing by big agribusinesses is pushing small-scale farmers closer to the wilderness and thus closer to wildlife pathogens.[25] The fact that humans are coming into ever closer contact with wildlife due to deforestation, urbanisation and expanding infrastructure is turning zoonoses into pandemics. In China, after agriculture intensified and was liberalised in the 1990s, small-scale farmers were driven to find new niches for their businesses and sell wildlife as a luxury food. This was publicly encouraged by the Chinese state, and the Wuhan wildlife market was integrated into the capitalist economy. In other words, Covid-19 started in China because China was *not* exotic or different.

Even deforestation must be mentioned here. This is an old activity, but comes in a new guise with modern capitalism. Deforestation is both

---

22 Rob Wallace, *Big Farms Make Big Flu: Dispatches on Influenza, Agribusiness, and the Nature of Science* (New York: Monthly Review Press, 2016); Rob Wallace, *Dead Epidemiologists: On the Origins of COVID-19* (New York: Monthly Review Press, 2020); Mike Davis, *The Monster at Our Door: The Global Threat of Avian Flu* (New York: New Press, 2005).

23 UNEP, *Emerging Issues of Environmental Concern*, United Nations Environment Programme Frontiers (Nairobi: UNEP Report, 2016).

24 Even if Covid-19 originated in a laboratory, it shows man's unhealthy relationship with nature and issues with the metabolic rift, and furthermore does not make the general analysis here any less relevant.

25 Rob Wallace et al., 'COVID-19 and Circuits of Capital', *Monthly Review*, 1 May 2020, monthlyreview.org.

the second most important cause of global warming (after fossil fuel emissions) and a direct cause of the creation of zoonoses and pandemics. So, what is behind the escalation of deforestation? Nowadays, it is giant capitalist monocultures, well integrated into global financial systems. And they produce beef, soybeans, palm oil and wood products – four products that account for about 40 per cent of tropical deforestation in Argentina, Bolivia, Brazil, Paraguay, Indonesia, Malaysia and Papua New Guinea.[26]

The UN states that the main causes of new diseases are land-use change (31 per cent), agricultural industry (15 per cent), international trade and travel (13 per cent), medical industry (11 per cent), war and famine (7 per cent) and climate change (6 per cent). The eating of bushmeat, which receives a disproportionate amount of attention, probably because hunting is often done by the poor, accounts for a modest 3 per cent.[27] Rather than exotic, new pandemics and other ecological crises are expressions of capitalism's latest advances. In the words of historian Peter Linebaugh, microparasites must be understood alongside macroparasites: to understand bacteria and viruses we must also comprehend the society and the economy.[28]

The Crisis of Man?

After 11,700 years of the Holocene we are at the doorstep of a new geological epoch, this time created by humans, hence the proposed name *Anthropocene*.[29] The argument is that human activity on the planet has left such a clear geological mark that even if all humans disappeared

26 Sabine Henders et al., 'Trading Forests: Land-Use Change and Carbon Emissions Embodied in Production and Exports of Forest-Risk Commodities', *Environmental Research Letters* 10, no. 12 (2015), pp. 1–14.

27 UNEP, *Emerging Issues*, p. 22. The remaining categories are human demography and behaviour (4 per cent), other (4 per cent), breakdown of public health (3 per cent) and food industry change (2 per cent).

28 'Peter Linebaugh on the Long History of Pandemics', 8 April 2020, in *Against the Grain*, podcast, kpfa.org.

29 For introductions, see, e.g., Lewis and Maslin, 'Defining the Anthropocene'; Angus, *Facing the Anthropocene*; Sörlin, *Antropocen*. On when the Anthropocene began, see, e.g., Steffen et al., 'The Anthropocene'; Waters et al., 'The Anthropocene'.

tomorrow, the human geological footprint will be clear millions of years into the future. Signs of our impact include climate change, biodiversity extinction, pollution, ocean acidity and the geochemical cycle of phosphorus and nitrogen. Some examples are just stunning: almost half of the earth's surface has been transformed by human actions; humans now move more sediment every year than all the winds and rivers combined; cities cover an area nine times larger than the whole of Sweden; enough plastic is produced each year to cover the entire planet in cling film; two-thirds of all arable land is farmed; and, between 1945 and 1963, an average of one nuclear test was carried out every ten days, which has left ineradicable traces on the planet.[30]

For socialists, the proposal of the 'Anthropocene' designation is ambiguous. On the one hand: we told you so! Capitalism has created a monster, and now even the geologists admit that the sorcerer – to paraphrase Marx – is no longer able to control the powers of the netherworld that he has called up by his spells.[31] But, on the other hand: An age of humankind? *Human*?

Firstly, many humans have not contributed to the problems at all. Between 1820 and 2010, population grew by a factor of 6.6, while CO_2 emissions grew by a factor of 654.8. Two billion people barely have access to electricity. Have they caused climate change?[32] A poor child dying at the age of ten for lack of food or medicine is just as much a human as the CEO of ExxonMobil. It would be preposterous to lump the two into one category and say that climate change is *their* fault. The Anthropocene might be a good term, Andreas Malm and Alf Hornborg note sarcastically, for the polar bear or the bird that wants to know which species is destroying their habitat. For those of us in the human kingdom, it is rather mystifying.[33]

The argument that it is humankind, *as humankind*, that causes the problems is something we also encounter in economic crises. As long as 200 years ago, chambers of commerce and the media described economic

30 Sörlin, *Antropocen*, pp. 32–3, 146.

31 Karl Marx and Friedrich Engels, *The Communist Manifesto* (London: Pluto Press, 2008), p. 41.

32 Andreas Malm and Alf Hornborg, 'The Geology of Mankind? A Critique of the Anthropocene Narrative', *Anthropocene Review* 1, no. 1 (2014), pp. 63–5; Angus, *Facing the Anthropocene*, pp. 112, 197.

33 Malm and Hornborg, 'The Geology of Mankind?', p. 67.

crises with strong moral overtones suggesting that they were caused, among other things, by human speculation and greed.[34] The explanatory model continues to pop up in economic crises. Humans *are* greedy; therefore they *must* create economic crises, just as we *must* consume the planet to death. A focus on humankind implies that it is our destiny to create these crises.

But human nature has always existed, and what has always existed can hardly explain anything specifically new here and now. This is, to paraphrase Andrew Kliman, like blaming a plane crash on gravity: gravity is always there, but planes do not always crash.[35] Explaining crises in terms of human nature removes power, society, economy, politics, complexity, class and colonialism – and, ironically, therefore, also actually existing humans – from the analysis.

A number of Marxists suggest that we should rather call the new geological epoch the *Capitalocene*. This points us in a different direction. Climate change has not been caused by humankind as an abstract figure, but by a specific political-economic system. Scientists who use the term Anthropocene, according to Jason Moore, reduce everything that happens to an 'abstract Humanity: a homogeneous acting unit'.[36] To which climate activist and eco-socialist Ian Angus rhetorically responds: Do proponents of Capitalocene really believe that the world's top climate scientists are so utterly stupid that they do not understand that poor people in Bangladesh are less responsible for climate change than billionaires in the US?[37]

The designation 'Capitalocene' is supposed to show that capitalism is the problem. Ironically, this term can also hide the central agents. Surely the working class is just as much part of capitalism and capital as the capitalist class? It would be more precise if we all started using 'Capitalist-class-ocene'. For good reasons, this will never happen. The term would also have to join a long queue of other proposals that have come up in recent years: Econocene, Misanthropocene, Chthulucene, Eurocene, Anthrobscene, Anthropo-obscene, White (M)Anthropocene, and so on

34 See, e.g., Reinhart Koselleck, 'Crisis', *Journal of the History of Ideas* 67, no. 2 (2006), pp. 389–90.

35 Andrew Kliman, *The Failure of Capitalist Production: Underlying Causes of the Great Recession* (London: Pluto Press, 2012).

36 Moore, *Capitalism in the Web*, p. 170.

37 Angus, *Facing the Anthropocene*, pp. 224–32.

and on. (Franciszek Chwałczyk identified in a paper from 2020 no fewer than ninety-one [*sic*!] different -ocenes.[38])

The concept of the Anthropocene – and, to a lesser extent, the Capitalocene – has burst into popular culture. Whether this really counts as a new geological epoch is ultimately decided by the International Commission on Stratigraphy (ICS) or the International Union of Geological Sciences (IUGS). For people actually affected by ecological crises, this formal decision matters relatively little. This is in contrast to the many academics who have invested heavily in the snazzy new concept. We will leave the question of a new geological age to the geologists. We shall stay here with an analytical framework I believe is both more precise and more politically relevant: the crises of capitalism.

But if I have to conclude – it is really hard not to – in the battle between the Capitalocene and the Anthropocene, I would say that yes, the Capitalocene is a far better concept. Should socialists then spend time and energy trying to convince the rest of the environmental movement about this? No. If anything, this shows how weak the left is. We are not even close to being able to stop the capitalist train heading straight into ecological disaster. But, along the way, we might be able to force a discussion about what to call the cliff.

But Is the Ecological Crisis Really a Crisis?

Not all disasters and problems on Earth should be defined as crises. Are the ecological disasters that surround us necessarily crises? In the introduction, we discussed four conditions for classifying phenomena as modern capitalist crises. The second condition was that the events should be grounded in underlying processes, the third was that they should have negative consequences and the fourth was that politicians or others must muster some kind of active response. All of this is clear with ecological crises and we will make it even clearer throughout the book. One condition remains, and that is the first: that crises are events that come relatively quickly.

The temporality of ecological crises is not entirely straightforward. Are they really events that come relatively quickly? First, if we take a specific

38 Franciszek Chwałczyk, 'Around the Anthropocene in Eighty Names: Considering the Urbanocene Proposition', *Sustainability* 12, no. 11 (2020), pp. 23–9.

hurricane, or Chernobyl, or the 2020 pandemic, then absolutely. But what about climate change, as the temporality here is less obvious? Humans have been burning coal under capitalism for over 200 years. For more than a hundred years scientists have known that the amount of carbon dioxide in the atmosphere affects temperature, and in recent decades the science has been very clear; anyone who does not actively choose to ignore reality knows this. While economic crises come as surprises that need to be studied after the fact, the climate crisis is simultaneously behind us, around us and in front of us. Here we can study tomorrow's crises today. Can we then say that the climate crisis is coming relatively quickly? Can it be called a shock? There are two ways to answer yes.

Firstly, while days and months may be a relatively short time for an economic crisis, years and decades are definitely a short time when we are talking about changing ecosystems that have been relatively stable for millennia. While it took hundreds of millions of years to form fossils, they have been burned up in a fraction of a geological second. In a few years, we must completely remake the global energy policy and eliminate the fuels that have created modern capitalism.[39] This is extremely urgent. From this perspective, the climate crisis is much more a shock than any economic crisis can ever be. The temporality of capitalism clashes with the timescale of evolutionary change. The climate crisis is very much the explosive meeting of capitalist market time and biological time.[40] One counter-argument here is that we humans do not really operate within geological time; we are subjugated by the temporality of capitalism. Which brings us to the second argument for why climate change is a shock.

Climate change is not a gradual and continuous process. Media often give the impression that the climate crisis has grown gradually over the course of several years and that it can be solved if we reduce emissions by so much over so many years. The climate crisis, in contrast, accelerates. Although systematic carbon dioxide emissions began 200 years ago, the great acceleration started only after the Second World War.[41] In 1950, the world as a whole emitted more than 5 billion tonnes of carbon

39 Angus, *Facing the Anthropocene*, p. 124.

40 See Michael Löwy, 'Daniel Bensaïd: A Marxism of Bifurcation', *International Viewpoint*, 28 June 2020; Saito, 'Marx's Theory of Metabolism', pp. 16–17.

41 See Steffen et al., 'The Anthropocene', p. 617.

dioxide, which is about the same as the annual emissions of the United States today.

Climate change also manifests itself as events. The hurricane that comes rushing in or the wildfires that threaten cities are not experienced as gradual processes. Ecological crises are becoming more numerous, more frequent and more severe. What really makes ecological crises rapid events and shocks is that their acceleration is not gradual: they come as abrupt changes.[42] According to Ian Angus, the conventional assumption is that the twenty-first century will be warmer and perhaps a little worse, but *not* fundamentally different. This assumption only works if we know exactly what is coming. We do not. For example, bacteria that have been frozen for millions of years may come to life when the permafrost thaws; ice melt could potentially open a Pandora's box of diseases we have no idea how to deal with.[43]

A concept from physics that has been established in climate research is *tipping points*.[44] The implications of this concept are so serious that they cannot be overstated. As the planet warms, what scientists call *feedback loops* are created. A classic example of a dangerous feedback system occurs when ice melts. Ice reflects away a lot of sunlight, but as ice melts, more and more sunlight is absorbed by the land and soil. This warms the planet further, which in turn causes more ice to melt, and so on in an increasingly vicious circle. Another example is that as the surface of water heats up, more water will evaporate. Since water vapour (H_2O) is a greenhouse gas, this will lead to further global warming, which, in turn, further warms the water which means more H_2O evaporates.[45] Some tipping points may come more abruptly (like large parts of the Amazon becoming savannah), while others may come a bit more slowly (like massive thawing of permafrost).

42 Will Steffen et al., 'Abrupt Changes: The Achilles' Heels of the Earth System', *Environment: Science and Policy for Sustainable Development* 46, no. 3 (2004), p. 9; see also Malm, *Corona, Climate*, pp. 16–19.

43 E.g., Jasmin Fox-Skelly, 'There Are Diseases Hidden in Ice and They Are Waking Up', BBC, 4 May 2017.

44 In relation to (Hegel's, Marx's and Engels's) discussions of quantitative and qualitative change, see Angus, *Facing the Anthropocene*, pp. 63–6.

45 One feedback system that – as an exception that proves the rule – can contribute to lowering the temperature is when more clouds counteract the warming of the planet. See Will Steffen et al., 'Trajectories of the Earth System in the Anthropocene', *Proceedings of the National Academy of Sciences* 115, no. 33 (2018), pp. 8254.

According to many of the world's leading scientists, continued global warming will take us to a point where strong non-linear feedback processes become more important than human emissions of greenhouse gases. Even if we then radically reduce emissions, climate change cannot be halted. The catastrophe is then beyond human control. Some argue that this limit is around two degrees Celsius of warming. It is hard to know exactly, partly because of the escalating nature of feedback and new mechanisms being discovered all the time, and partly because these feedback systems interact and reinforce each other.[46] If we lose control, scientists say it could potentially lead to a planetary situation last seen millions of years ago, making it impossible to reproduce our current human societies.[47]

Ecological and economic crises have both similar and different rhythms. Nancy Fraser emphasises the former, and places the crises within an identical historical framework in which they represent breaks between four historical epochs of capitalism: mercantile capitalism, liberal competitive capitalism, state-led capitalism and finance capitalism. At the epochal shifts, the economy, ecology, racism and social reproduction all change in the same direction, showing that the economy and political ecology are embedded in the same system.[48] But economic and ecological crises also have different temporalities. Where economic crises come and go, ecological crises seem to mainly just come. They will come more often and be more and more serious. The economic crisis – as we saw in chapter 1 – is at the same time both the crisis itself and the solution to the crisis. The ecological crisis, on the other hand, is simultaneously both a crisis and the potential demise of our societies as we know them.

We do not know exactly how the climate crisis will unfold, but we know very precisely where it comes from. Compared to the complex

---

46 Dan Lashof, 'Why Positive Climate Feedbacks Are So Bad', World Resource Institute, 20 August 2018; James Hansen et al., 'Target Atmospheric CO_2: Where Should Humanity Aim?', *Open Atmospheric Science Journal* 2 (2008), pp. 217–31.

47 Steffen et al., 'Trajectories of the Earth System', p. 8253.

48 Nancy Fraser, in Nancy Fraser and Rahel Jaeggi, *Capitalism: A Conversation in Critical Theory* (Cambridge: Polity, 2018), pp. 96–101. See also Andreas Malm, 'Long Waves of Fossil Development: Periodizing Energy and Capital', *Mediations* 32, no. 1 (2018), pp. 17–40.

causes of economic crises – where there are several factors we will never fully understand – climate change is fairly simple. In contrast to how economists will never agree on the main causes for economic crises, most scientists do agree on the main causes for climate change. Although ecology and climate are highly complex systems, the cause of global warming is basically quite straightforward: if we emit more greenhouse gases, it gets warmer.

Although the causes of global warming are crystal clear, it seems almost impossible to actually solve the problem. Again, we see a striking contrast with economic crises. With economic crisis, causes are complex, but we can be confident that the world's leaders and the ruling class will always (try to) solve the crisis. With global warming, we know the causes, but the world's leaders and the ruling class have seemingly no idea how to stop it. Why does the easy problem seem so hard to solve, while the hard problem is so simple? This may seem illogical, but, in fact, it is not.

The answer lies precisely in the fact that these are two different capitalist crises: they are both *capitalist* crises – rooted in the same political economy – but also *different*. As events, they are surely distinct phenomena. But they are also different in the sense that the same underlying processes that constitute them have different impacts on our two crises. The same general mechanisms and structures within the capitalist system make it easier to solve economic crises but harder to solve the ecological ones.

With economic crises comes creative destruction, something partly created by the logic of capital accumulation, and which contributes to solving the crises. In addition, the ruling class and leading politicians will do *whatever it takes* to solve economic crises, as this is decisive for their own reproduction as rulers of the world. Their actions are very different in the case of ecological crises. Indeed, it is harder to stop global warming *because of* the logic of capital accumulation. We will return to the issues of class, racism and nationalism later in the book; in the rest of this chapter, we will just discuss some of these aspects of capitalism that make it difficult to solve ecological crises.

Not Solving the Climate Crisis

Solving the climate crisis under capitalism, it is often said, is like walking up an escalator that is rolling towards you – at an accelerating pace. So let us get to the bottom of exponential growth. The capitalist economy does not grow by a fixed amount every year, it grows by a certain percentage. Which makes an incredible difference. The crux is that the economy grows more in the second year than the first, and even more in the third year than the second. At a growth rate of 2.1 per cent each year, the world's total output will be eight times larger in 100 years, and sixty-four times larger in 200 years. As capital accumulation is an unstoppable process, growth will increase forever. Or until we change economic systems.

The relationship between capital accumulation, profit and economic growth is not straightforward. It is never one-to-one: a certain amount of profit does not directly create a corresponding amount of economic growth. The relation has also varied historically: in the 1960s and '70s, growth was still relatively good while profits were falling; under neoliberalism, growth in the Global North was significantly lower, while profits in many sectors were relatively strong. Relations between profit and growth are never one-to-one, *but there are relations*.

Debates over 'growth' are often confusing, not least since the concept is used in very different ways. When people discuss 'for or against' growth, do we mean growth in the use of biophysical or material throughput, in energy use, in human potentials, in capital accumulation, in PPP (purchasing power parity) or HDI (human development index) as a purely metaphysical idea, or is growth merely an ideology? Here, we will focus on economic growth, which is conventionally measured by gross domestic product (GDP), a measure of all economic activity in a country in a year. But there are even different methods of measuring GDP and also other ways of measuring the economy.[49] GDP has been criticised for various reasons. Increased production of weapons and private jets can increase GDP while being mostly destructive for the human race. Critical scholars argue, therefore, that we need new ways of measuring economic activity, focusing on human development

49 For discussions, see Rikard Hjorth Warlenius, *Klimatet, tillväxten och kapitalismen* (Stockholm: Verbal, 2022), ch. 3.

or happiness.[50] A new measurement may be important, but the key point is not how we think about the issue or how we measure growth. It is that we de facto have an economy that *necessarily* expands and needs increased input of materials of various sorts.

If one accepts the premise that profit is the primary motive behind capital accumulation, does this necessarily imply economic growth? One can imagine various hypothetical scenarios, such as all surplus value going to wages rather than profits (but this clashes with capital's need for accumulation), producing smarter rather than more (but increased efficiency has never replaced quantity so far under capitalism), or politicians in one country banning profits (but then firms could simply be out-competed by firms in other regions). One tradition that argues that capitalism is, by definition, *not* tied to growth is steady state economics. This is a (theoretical) variant of capitalism according to which population and physical stock/wealth – and hence use of natural resources – do not increase, but the economy still develops technologically and ethically.[51] The tradition can be traced back to Adam Smith and John Stuart Mill, and today someone like the British economist Ann Pettifor advocates steady state economics in her manifesto for a Green New Deal. Here, credit and not profit is the main driver of economic expansion, and the concept of capitalism is almost non-existent.[52]

If capitalist growth is a bad thing, what about capitalist non-growth? The case of Japan is often used today as proof that capitalism can be reproduced without the economy needing to grow. However, Japan experienced low economic growth in a *world* of growth, and many Japanese companies survived by investing abroad. Many countries in Latin America experienced zero growth during, for example, the debt crisis in the 1980s, but this should not be considered a steady state economy as their crisis contributed to growth elsewhere. That individual countries

50 See, e.g., Kate Raworth, *Doughnut Economics: Seven Ways to Think Like a 21st-Century Economist* (White River Junction: Chelsea Green Publishing, 2017), ch. 1.

51 Ann Pettifor, *The Case for The Green New Deal* (London: Verso, 2020), chs 4–5; Lars Pålsson Syll, *De ekonomiska teoriernas historia* (Lund: Studentlitteratur, 2007), pp. 143–4; for criticism, see Malm, *Fossil Capital*, pp. 284–5; Frederik Berend Blauwhof, 'Overcoming Accumulation: Is a Capitalist Steady-State Economy Possible?', *Ecological Economics* 84 (2012), pp. 254–61; Richard Smith, 'Beyond Growth or Beyond Capitalism?', *Real World Economics Review* 53 (2010), pp. 28–36; see also Warlenius, 'Inledning', pp. 16–21; Warlenius, *Klimatet*, pp. 54–9, 329–37.

52 Pettifor, *The Case.*

can survive without economic growth is not unusual, especially in crises, but that capitalism as a whole could function over time without growing at all is another matter altogether. It should also be mentioned that Herman Daly, who coined 'steady state economics', later argued that for him this was not the same thing as 'steady state capitalism'.[53]

The desire to take growth out of capitalism reminds me of what Paul Sweezy once said about Keynes's ideas: that they tear the economic system out of its social context and treat it as if it were a machine to be sent to the repair shop, there to be overhauled by an engineer state.[54] Steady state economics is a variant of the old liberal dream of removing the problems of capitalism and keeping the system itself. Proponents have been criticised for seeing capital as a constant quantity or accumulation of physical wealth. If we understand capital as a process – as value in motion – rather than as a finite and static thing, accumulation can never be taken out of its social context.

There are also relations between capital accumulation and economic growth, on the one hand, and environmental impact, on the other. And again we can say that the relation is never one-to-one, but there are indeed relations. The exploitation of nature in a growing economy is based on the same principle as capital accumulation. New rounds build on the previous ones. If capital grows from 100 to 1,000, Marx explains pedagogically in *Grundrisse*, 1,000 is now the new starting point. The fact that the economy has multiplied does not really matter. What were previously surplus value and profit now become given conditions for new turns of accumulation.[55] The same applies to nature. The more biophysical resources are used to create profit and growth, the more can and must be extracted accordingly.[56] Today's mistreatment of nature is based on the unsustainable use of nature over centuries. Tomorrow's cycles of capital accumulation are built on generations of exploited labour and destroyed nature.

The phenomenon of growth is so obvious and so explosive that our leaders cannot avoid it. It was central to both of the most important and

53 See Warlenius, *Klimatet*, p. 330.

54 Paul Sweezy, *The Theory of Capitalist Development* (London: Dennis Dobson, 1946), p. 349.

55 Karl Marx, *Grundrisse* (New York: Vintage, 1973), p. 335; Marx, *Capital*, vol. 1, p. 729.

56 See also Malm, *Fossil Capital*.

best-known environmental reports of the twentieth century. *The Limits to Growth* was written on the initiative of the Club of Rome and published in 1972, and *Our Common Future*, often referred to as the Brundtland Report, was written on behalf of the UN and published in 1987. Fifteen years separate these two environmental reports, but this was not just any fifteen years. Between the reports, the world was revolutionised. The reports were written on either side of the economic crisis of the 1970s, the oil shocks of 1973 and 1979, and one before and the other after the fall of Keynesianism and the triumph of neoliberalism. They seem to come from different planets. The linguistic focus on growth in *The Limits to Growth* was often replaced by a focus on development in *Our Common Future*. In terms of content, the former considered growth to be a massive problem; the latter saw it as a prerequisite for achieving sustainable development.[57]

The economic historian Jenny Andersson argues that *The Limits to Growth*, as well as 'Third World' countries becoming increasingly independent and self-reliant, prompted an organisation like the OECD to develop new discourses and policies in the 1970s that envisioned a different form of globalisation. Andersson calls it 'proto-neoliberalism'.[58] It was simply not appropriate for the capitalist class or the world's richest countries to put any restrictions on capitalist development because some scientists had concluded that nature had certain limits.

Since *Our Common Future*, the world's leaders have never looked back. The hegemonic view of environmental crises has become ecological modernism and green capitalism. Here, solutions to environmental problems – one way or the other – can be found within the framework of modern capitalism.[59] Other views surely exist, one being climate

57 Donella H. Meadows et al., *The Limits to Growth*, commissioned by the Club of Rome (Washington, DC: Potomac Associates, 1972); World Commission on Environment and Development, *Our Common Future*, Report of the World Commission on Environment and Development (1987).

58 Jenny Andersson, 'The Future of the Western World: The OECD and the Interfutures Project', *Journal of Global History* 14, no. 1 (2019), pp. 126–44.

59 See Paul McLaughlin, 'Ecological Modernization in Evolutionary Perspective', *Organization and Environment* 25, no. 2 (2012), pp. 178–96; Arthur P. J. Mol, 'Ecological Modernization as a Social Theory of Environmental Reform', in Michael R. Redclift and Graham Woodgate (eds), *The International Handbook of Environmental Sociology* (Cheltenham: Edward Elgar, 2010), pp. 63–76; Richard York, Eugene Rosa and Thomas Dietz, 'Ecological Modernization Theory: Theoretical and Empirical Challenges', in

denialism: people who either *do not believe* that the world is getting warmer or that humans are responsible, or *do believe* that a warmer world is good for humans overall.[60] Another group is the 'mad professors' who have worked out that climate change is worth it. The statistician Bjørn Lomborg concluded that it was not worth doing anything about the problems at all, and two economists in the *Wall Street Journal* wrote that climate change cannot justify or legitimise policies that cost more than 0.1 per cent growth.[61] In addition to ignoring tipping points and class analysis, these gentlemen show utter contempt for affected people and for the earth. For those who actually die, it matters less if some Danish statistician has calculated that their death is profitable for someone else. But the hegemonic view continues to be eco-modernism: the dream of a green capitalism where new technologies, green consumption, more economic growth and ecological improvements can go hand in hand into eternity. This immediately solves several problems for both the capitalist class and politicians: the former retains and legitimises its social power, the latter can promise their allies and voters more of everything – higher profits, higher wages, more welfare, bigger budgets *and* a better environment.

A common hypothesis for ecological modernists is that the economy can be dematerialised.[62] Endless growth would, of course, be less of an ecological problem if it were just a matter of numbers in a bank account, financial bubbles or inflation. Perhaps services, education, culture and experiences can pull the entire global economy forward? Eternal capitalism without climate impact! The 'post-industrialisation' of the Global North was always a geographical relocation of industrial production.

Redclift and Woodgate, *The International Handbook of Environmental Sociology*, pp. 77–90; Richard York and Eugene Rosa, 'Key Challenges to Ecological Modernization Theory', *Organization and Environment* 16, no. 3 (2003), pp. 273–88; Petter Næss, 'The Illusion of Green Capitalism', in Petter Næss and L. Price (eds), *Crisis System: A Critical Realist and Environmental Critique of Economics and the Economy* (London: Routledge, 2016), pp. 173–91; Foster, Clark and York, *The Ecological Rift*, pp. 41–5. Recently, the terminology has become somewhat problematic, as even socialists are now identifying as eco-modernists.

60 For an overview, see Malm and the Zetkin Collective, *White Skin, Black Fuel*.

61 Bjørn Lomborg, *The Skeptical Environmentalist* (Cambridge: Cambridge University Press, 2001); David R. Henderson and John H. Cochrane, 'Climate Change Isn't the End of the World', *Wall Street Journal*, 30 July 2017.

62 Petter Næss, 'Unsustainable Growth, Unsustainable Capitalism', *Journal of Critical Realism* 5, no. 2 (2006), p. 200.

(As we know, headquarters and ownership seldom moved to the Global South. If they moved at all, it was rather to tax havens.) In stark contrast to what many believe, the global workforce in the industrial sectors is growing. Between 1991 and 2012, during the heyday of so-called post-industrialisation, the number of workers in goods manufacturing and industry globally increased from 490 million to 715 million people. This sector actually grew *faster* than the service sectors.[63]

Capital accumulation and thus a growing economy come with increased production of physical goods. Historically, there have been more buildings and cities and bigger airports and ports, and there has been an increase in infrastructure, plastic, paper, raw materials and minerals, all requiring more and more energy. Should the global post-industrial society miraculously materialise, it would mean a *whole new form* of capitalism. To stop global warming, this needs to be implemented within a few years. This requires, I would say, a revolution more radical and much less realistic – since it is not even theoretically clear as to how it could be achieved – than eco-socialism.

Everything is not getting worse everywhere all the time. At different places and in different countries, different improvements have been made, like industries getting cleaner, city centres becoming car-free and so on. These must be strongly defended.[64] Because of such improvements, even if they are limited in time and space, ecological modernists are affected by hubris. The environmental Kuznets curve is a hypothesis that says, while growth negatively affects the environment in the early development of a country's economy, after a certain point there is a reversal: thereafter, increased growth will lead to *fewer* environmental problems. Again, this is true in some sectors, in some places, and for some time, although it is certainly more about policy regulations than growth itself. Often, the problems have simply been exported to poorer countries. For the theory – assuming it is correct – to work on a global scale, *all* countries must develop and become rich. This is not the case in an imperialist world where dominant interests benefit from underdevelopment and poverty.

63 Michael Roberts, 'Deindustrialisation and Socialism', *The Next Recession*, 21 October 2014.

64 Cf. Brett Clark and John Bellamy Foster, 'The Environmental Conditions of the Working Class: An Introduction to Selections from Frederick Engels's *The Condition of the Working Class in England in 1844*', *Organization and Environment* 19, no. 3 (2006), p. 387.

There is plenty of data, then, to suggest that the environmental Kuznets curve is simply wrong.[65] It fits poorly with reality, but well with the way leading politicians in the Global North approach these problems: only with high growth and a good economy can we afford to think about ecological issues.

If continued growth and further accumulation improve the environment, one could think that absence of economic crises would lead to less climate change and a better environment. When looking at actual emissions, we see that the curve points in the exact opposite direction. Historically, carbon dioxide emissions have always increased over the span of modern capitalism, with the exception of 1929, 1944, 1973, 1979, 1989, 1991, 2008 and 2020. If we disregard 1944 (Second World War) and 1989 (the collapse of the Eastern Bloc), the common denominator for years with lower emissions is economic crisis.[66] Somewhere here, a curious phenomenon emerges. While politicians become less interested in ecological crises in times of economic crisis, it is almost exclusively during such crises that greenhouse gas emissions actually fall. Emissions only fall when the ruling class *does not* try to do anything about them.

Profits are simultaneously causing the climate crisis, a reason we cannot stop it, and a premise for attempting to deal with it.

Insufficient Decoupling

As the history of capitalism is largely marked by technological innovation, efficiency and increased productivity, it is perhaps not surprising that many are staking their prestige – and our future – on finding the solution to environmental problems among these phenomena. The key is to *decouple* growth from increased environmental pressure. The

65 Warlenius, *Klimatet*, pp. 241–4. For research reviews that debunk the environmental Kuznets curve, see Goodness C. Aye and Prosper Ebruvwiyo Edoja, 'Effect of Economic Growth on CO_2 Emission in Developing Countries: Evidence from a Dynamic Panel Threshold Model', *Cogent Economics and Finance* 5, no. 1 (2017); Helmut Haberl et al., 'A Systematic Review of the Evidence on Decoupling of GDP, Resource Use and GHG Emissions, Part II: Synthesizing the Insights', *Environmental Research Letters* 15, no. 6 (2020).

66 On data up to 2008, see Climate Watch Data, 'Global Historical Emissions', 2020, climatewatchdata.org. For 2020, see Zhu Liu et al., 'Near-Real-Time Monitoring of Global CO_2 Emissions Reveals the Effects of the COVID-19 Pandemic', *Nature Communications* 11 (2020), p. 1.

concept of *relative decoupling* refers then to a reduction in environmental impact – or carbon intensity, if you like – per unit produced.[67]

Let us begin by recognising that technological changes that make production more sustainable and energy-efficient are needed. This is undoubtedly necessary if we are to solve the climate crisis. Research and policy change that contribute to a smaller environmental impact are welcome. A relative decoupling has also taken place in recent decades, at least in some sectors in some countries for some time. This is all positive.

But then come the problems. And they go right to the heart of capitalism. In context of climate change, it does not matter that every commodity can be made to be more environmentally friendly if more and more commodities are produced. It is little consolation that production will be *relatively* better if *absolute* emissions increase. Here is the crux. For relative decoupling to lead to actual emissions reductions, relative improvements must be greater than economic growth. In addition, the absolute decoupling must also be a *sufficient decoupling* in order to reach climate or other environmental targets.[68] Since capital accumulation causes growth to increase exponentially, the decoupling must increase with ever greater intensity. Since capitalism goes on and on, relative improvement must be better than growth in perpetuity – or for as long as capitalism exists. Decoupling under feudalism – or other economies not creating perpetual exponential economic growth – would have quite different consequences.

Human ecologist Helmut Haberl and his colleagues have analysed no fewer than 835 research articles on the subject and conclude that absolute decoupling occurs only in exceptional cases, within individual countries, if the focus is on production-based rather than consumption-based emissions figures, and preferably in conjunction with periods of low economic growth. The vast majority of studies point in the opposite direction.[69] It is, therefore, not enough to urge consumers to be environmentally friendly or companies to invest in renewable energy, Haberl and colleagues conclude: the non-renewable must be banned.

67 Tim Jackson, *Prosperity without Growth: Foundations for the Economy of Tomorrow* (London: Routledge, 2016).

68 Warlenius, *Klimatet*, pp. 234–41.

69 Haberl et al., 'A Systematic Review', pp. 31–4.

Not only is relative decoupling inadequate; it can also be part of the problem. Stanley Jevons famously showed in *The Coal Question* of 1865 that, as steam engines became more energy efficient and used less coal, the demand for coal did not decrease, but rather increased. Let us dust off the old Jevons paradox with a modern example.[70] If an airline uses an engine that requires less fuel, emissions per flight will decrease. This is relative decoupling, and generally a good thing. If the motive of the company were to fly a certain number of people with the least possible emissions, this would lead to a real improvement. But the motive under capitalism is profit, and whether the company implements a new engine will ultimately be determined by profit. What will the company do with the increased profit from using less fuel? In order to keep up with the perpetual competition, the company would have to put large parts of the capital into new rounds of accumulation. Which would mean, since we are talking about an airline company, more planes and more departures. In other words, more relative decoupling can lead to more profit and growth, and thus to *less* absolute decoupling.

A growing economy creates a number of challenges. For example, new investment in renewable energy does not directly replace existing fossil fuel production. It might just add to overall energy production.[71] Moreover, the boundaries within capitalism between the so-called green, grey or black sectors of the economy are unclear. If you spend a hundred euros on trains instead of flying, you cannot know how the owner – who may be a corporation headquartered on a Caribbean island – reinvests this money. What is the alternative? Not spending the money at all and saving it in the bank? Hardly. In such a case, the bank will lend your money for investment somewhere else.

We live in a type of capitalism that runs on fossils, but could we have lived in a different capitalism? Could capitalism function without fossil fuels?

70 W. Stanley Jevons, 'The Periodicity of Commercial Crises, and Its Physical Explanation', *Journal of the Statistical and Social Inquiry Society of Ireland* 7, no. 54 (1878–79), pp. 334–42. The modern example is stolen from Warlenius, 'Inledning'. See also Armon Rezai and Sigrid Stagl, *Ecological Macroeconomics: Introduction and Review*, Working Paper Series No. 9, 2016, Institute of Ecological Economics, Vienna University of Economics and Business, p. 2.

71 Cf. Richard York, 'Do Alternative Energy Sources Displace Fossil Fuels?', *Nature Climate Change* 2 (2012), pp. 441–3.

The historical answer is yes. The answer is historical because capitalism has done just that: it existed before it started using fossil fuels. Could capitalism have grown so explosively without fossil fuels? No. It was perhaps not a given that capitalism would create a new geological epoch, but capitalism as we know it is unthinkable without fossil fuels. If humans had never integrated fossil fuels into the economy, would we live in a type of capitalism that has a sustainable relationship with nature? The answer is no. A system based on the perpetual pursuit of profit, exponential growth and on a metabolic rift has in its basic structure an unsustainable relationship with nature. *Capitalism can never be sustainable.*

Mechanisms in capitalism make it more difficult to solve ecological crises, but does this mean that nothing is possible under capitalism? First, we must be careful never to underestimate how flexible capitalism is. We also need to emphasise how radical shifts in technological paradigms can be. For example, even within a fossil-driven capitalism, Sweden's carbon dioxide emissions fell by 35 per cent between 1976 and 1983, to a large extent because of the massive expansion of district heating and nuclear power.[72] Again, it is specific to a place and a time, and nuclear power certainly comes with its own challenges, but 35 per cent in seven years remains noteworthy. (This transformation has also been used as an argument that crises are indeed 'opportunities' for environmental movements, but then we must add that this was mainly a response to questions of national energy and security, not progressive environmentalism defeating powerful fractions of capital.)

It does matter what energy communities run on and how energy systems are organised, even within capitalism. The question that follows from this is whether a similar paradigm shift could happen on a global scale – a world run on renewable energy – even within the framework of capitalism. Theoretically, yes. Could we, despite the processes and challenges described above, stop temperatures from increasing more than two degrees? Perhaps. Capitalism is indeed a form of production marked by plasticity and it is certainly possible to imagine a ban on fossil fuels, massive financing of renewable energy, and a historic expansion of renewable energy sources and nuclear power – even within the framework of capitalism. Is it realistic? I suggest it is not. I will not pretend to be less ambivalent. I am, however, very sure of four other things.

72 See, e.g., Warlenius, *Klimatet*.

First, I know that this will not happen without class struggle. Fossil capital will never surrender voluntarily, and can only be broken down through struggle. Within capitalism, this means that fossil capital must be crushed by other elements of capital – along with the environmental and labour movements. Second, this will be impossible without state power, not only to develop non-fossil energy sources, but – far more importantly – to actively ban the fossil sector. Third, if the rise in temperatures is kept below 1.5 or 2 degrees for a few decades within the framework of capitalism, the problems will not go away. Economic growth requires the constant building of more and more renewable energy, and the extraction of more and more minerals. This will come at an enormous cost. Many other ecological problems will become increasingly difficult to solve, as 'green capitalism' requires huge amounts of minerals and resources, new machinery and infrastructure, and so on. Fourth, it would be infinitely easier to solve the climate crisis if we had a planned socialist economy that did not depend on perpetual exponential growth.

In the face of climate change, we need to keep two thoughts in mind at the same time. Some goods, sectors, technologies and so on are better than others, but, at the same time, profit motive, exponential growth and private ownership – *capitalism itself* – make the crisis harder to solve. In other words, it is very difficult to stop global warming precisely because it is a capitalist crisis.

That ecological crises are capitalist crises is not something we learn from the morning paper or on TV. When the media are presented as being neutral, it usually means they have placed themselves in the middle ground between different people and positions that all take capitalism for granted.

We are told that we need more education and knowledge, but we already know more than enough. A classic myth is that the climate crisis is escalating because of general ignorance about climate change. Ironically, the historical trend has been in the opposite direction. Emissions have been escalating while we have been gaining more and more knowledge about the subject.[73] The climate crisis is not about knowledge, but about power.

73 Kari Marie Norgaard, *Living in Denial: Climate Change, Emotions, and Everyday Life* (Cambridge, MA: MIT Press, 2011); Malm and the Zetkin Collective, *White Skin, Black Fuel*; Matt Huber, *Climate Change as Class War: Building Socialism on a Warming Planet* (London: Verso, 2022).

We are told that ordinary people are the problem, but in reality it is the ruling class that is the main issue. The focus on recycling, sustainable shopping, doing this and not that leads one to think that we – ordinary people on planet Tellus – have caused our problems through unsustainable consumption. Consumption as a social phenomenon is important for a number of reasons, but questions of power disappear when the focus is on individual consumption within the working class. If we are to place responsibility anywhere, we must place it with those in power: those who actually own oil companies, airlines and car companies, as well as politicians who have approved the climate crisis for decades.

We are constantly told that poverty is the problem, but, in fact, it is excessive wealth that is our great concern. We are also told that overpopulation is a problem because too many children are being born in India and Nigeria. Conversely, we are never told that too many children are born in Danderyd or Östermalm – the bourgeois parts of Stockholm. You are not allowed to say that. This is taboo because we live in a world where the rulers and the rich have platforms to offer opinions about the propensity of the poor to reproduce, but where the poor could never offer opinions about the rich.

We are told that we need to build more green energy, but, for stopping global warming, the key is how much fossil energy we *remove*. We are told that the climate crisis will be solved when environmentally friendly alternatives become cheaper on the market. But it is not the price that matters, it is the profit rate. The great optimist points out that green energy is getting cheaper and cheaper and that so many new solar panels are being built. The oil industry will not stop oil production because wind power is cheaper. It will stop producing oil when it is no longer profitable to do so. And not necessarily even then, because of its geopolitical interests. Oil must be banned.

Still, the news is full of our two crises. Economic and ecological problems fill our TV screens daily. Despite the fact that news broadcast after news broadcast is devoted to our crises, we rarely hear anything sensible about how they are actually related to each other – or to capitalism.

3
Rooted in Capitalism

Crises are social paroxysms. They are, by definition, combinations of general features (fundamental contradictions in the capitalist mode of production) and specific events (arising from the historical moment in the development of capitalism when the crisis occurs). The further we probe into the crises, the more the particular stories and details of individual crises disappear, but, at the same time, new aspects and contradictions develop. If we want to understand the relationship between ecological and economic crises, how crises both shake and stabilise capitalism, and how crises appear as weaknesses but are strengths for the capitalist system, then we cannot avoid examining both the general and the particular.

If we see each crisis only as a unique event, we miss why they recur and the context – that crises are always crises within something. Then we go down in history as amnesiacs. On the other hand, if we see crises only as predetermined expressions of underlying processes, we miss the specificity of each crisis; we miss the people who influence its development and the real nature of crises as defining moments. Then we are classed as politically irrelevant armchair theorists. Underlying processes show why crises are relatively similar and recurrent, while more superficial phenomena point to how they are relatively different and how they are experienced as new each time. What makes crises crises, what distinguishes them from random catastrophes or permanently ongoing processes, is precisely that they exist in a dialectical relationship between surface and underlying phenomena, between the particular and the general – that is, between different levels of abstraction.[1]

Analysing crises as both concrete events and underlying contradictions is an intellectual challenge: to constantly understand the big in the small and the small in the big – events in context, context in events – without reducing everything to one or the other. When we talk about crises, we simply cannot duck the challenge. It must be confronted.

Five Levels of Abstraction

Abstraction is about creating shortcuts, ignoring the complexity of reality and making things simpler than they really are.[2] Hegel argued that

1 On economic crises and levels of abstraction, see, e.g., David Harvey, *The Enigma of Capital: And the Crises of Capitalism* (London: Profile Books, 2010); David Harvey, 'History versus Theory: A Commentary on Marx's Method in Capital', *Historical Materialism* 20, no. 2 (2012), pp. 3–38; Bob Jessop, 'The Symptomatology of Crises, Reading Crises and Learning from Them: Some Critical Realist Reflections', *Journal of Critical Realism* 14, no. 3 (2015); Annika Bergman-Rosamond et al., 'The Case for Interdisciplinary Crisis Studies', *Global Discourse: An Interdisciplinary Journal of Current Affairs* 12, no. 3–4 (2022), pp. 465–86; Anwar Shaikh, 'An Introduction to the History of Crisis Theories', in Union for Radical Political Economics (ed.), *U.S. Capitalism in Crisis* (New York: Union for Radical Political Economics, 1978), p. 219; Rick Kuhn, 'Henryk Grossman and the Recovery of Marxism', *Historical Materialism* 13, no. 3 (2005), p. 9; Henri Lefebvre, *The Production of Space* (Oxford: Basil Blackwell, 1991).

2 Bertell Ollman, *Dance of the Dialectic: Steps in Marx's Method* (Chicago: University of Illinois Press, 2003), pp. 61–2. On abstractions see, e.g., Lefebvre, *The Production*; Henryk Grossman, 'Marx, Classical Economics, and the Problem of Dynamics', *International Journal of Political Economy* 36, no. 2 (Summer 2007), pp. 6–83; David Harvey, *Justice, Nature, and the Geography of Difference* (Cambridge, MA: Blackwell, 1996); G. W. F. Hegel, *Who Thinks Abstractly?* (Garden City, NY: Anchor Books, 1966), pp. 113–18; Don Mitchell, 'There's No Such Thing as Culture: Towards a Reconceptualization of the Idea of Culture in Geography', *Transactions of the Institute of British Geographers* 20, no. 1 (1995), pp. 102–16; Andrew Sayer, *Method in Social Science: A Realist Approach* (London: Routledge, 1992); David McNally, 'Beyond the False Infinity of Capital: Dialectics and Self-Mediation in Marx's Theory of Freedom', in Albritton Robert and John Simoulidis (eds), *New Dialectics and Political Economy* (London: Palgrave, 2003); Karl Marx, *Capital*, vol. 1 (London: Penguin, 1976); Lukasz Stanek, 'Space as Concrete Abstraction: Hegel, Marx and Modern Urbanism', in Kanishka Goonewardena et al. (eds), *Space, Difference, Everyday Life: Reading Henri Lefebvre* (New York: Routledge, 2008), pp. 62–80; Alberto Toscano, 'The Open Secret of Real Abstraction', *Rethinking Marxism* 20, no. 2 (2008), pp. 273–87; Paul Sweezy, *The Theory of Capitalist Development* (London: Dennis Dobson, 1946), pp. 133–4.

abstractions were poor and one-sided while the concrete was complex and difficult; Henri Lefebvre argued that abstractions do violence to and destroy reality. *Abstrahere* (from Latin) also means roughly 'to take out', 'to disregard', 'to omit details'; for Hegel, it means, roughly, to separate. Abstractions are simplifications we need in order to discuss anything at all.[3] For example, there are unclear boundaries between our two crises: the economy is central to ecology and vice versa, and no one can point to the exact boundary between society and nature. Despite a series of unclear boundaries, it does not make sense to insist that everything is *one* crisis.[4] The mess becomes impenetrable if we do not *separate* between levels of abstractions, and between our two crises.

The concept of abstraction is used in different ways, as it will also be used in this chapter. We begin with abstractions as a thought process and method. The American political scientist Bertell Ollman provides us with three concrete tools here: extension, levels of generality and vantage points.[5]

With *extension*, we set temporary and spatial boundaries. The world is constantly changing, but we have to pretend that it is standing still. To talk about specific places in a globalised world, we have to pretend that the rest of the world does not exist. When we discuss a crisis, there are millions of aspects that can be included. Not all of them are equally important by any means, but which ones and how many we include is a question of extension.

The second tool to make abstraction is the *level of generality.* For example, we can see the Great Depression of 1929 as 1) a stock market crash, 2) an economic crisis, which was part of a broader 3) organic crisis in capitalism, which is a phenomenon in 4) human evolution, which, in turn, is part of 5) the animal kingdom, and 6) nature.

---

3 See, e.g., Ollman, *Dance of the Dialectic*, p. 60; McNally, 'Beyond the False Infinity', p. 8; Lefebvre, *The Production*, p. 289; David Harvey, *A Companion to Marx's Capital* (London: Verso, 2010), p. 16.

4 For a defence, see Jason W. Moore, 'Toward a Singular Metabolism: Epistemic Rifts and Environment-Making in the Capitalist World-Ecology', *New Geographies* 6 (2014), p. 13; Jason W. Moore, 'The End of Cheap Nature: Or How I Learned to Stop Worrying about "The" Environment and Love the Crisis of Capitalism', in C. Suter and C. Chase-Dunn (eds), *Structures of the World Political Economy and the Future of Global Conflict and Cooperation* (Berlin: Lit Verlag, 2014), pp. 288, 290. For a critique see John Bellamy Foster, 'The Epochal Crisis', *Monthly Review* 65, no. 5 (2013), p. 1; Kamran Nayeri, '"Capitalism in the Web of Life": A Critique', *Climate and Capitalism*, 19 July 2016.

5 Ollman, *Dance of the Dialectic*, ch. 5.

Vantage points comprise the third tool. They colour what we see, create order, hierarchies and priorities, rank meanings, and highlight certain relationships at the expense of others. Had we examined the phenomenon of crisis from the point of view of racism or gender, both the crises and capitalism would have appeared different. Not necessarily in ways that are mutually exclusive from our analysis – on the contrary, good analyses strengthen and improve each other – but other aspects would have emerged.

Everyone has to make abstractions, but not all abstractions are equally good. Human geographer Don Mitchell argues that abstractions can be good (i.e., rational and concrete) or bad (chaotic and either too big or too limited). What matters is how and what we abstract, and that we are aware of what we are doing. Some abstractions can help us reveal power relations, while others can hide them. Useful abstractions are, according to Mitchell, 'firmly rooted in specifiable processes and denote an internal coherence'.[6] Our challenge is to first capture the complex reality and then present it in the most accessible way possible.

So far in this book, we have used terms such as 'surface' and 'underlying' phenomenon rather imprecisely. In what follows, we will make the analysis both more complex, but, we hope, also more comprehensive, by examining the economic and ecological crises at five levels of abstraction:

1) the shock and the event
2) the concrete organisation of economy and nature
3) crisis tendencies inherent in capitalism
4) the eternal pursuit of profit
5) use-value versus exchange-value

It is an analytical challenge to move between levels of abstraction, and there is never one tendency or one level that directly determines everything else. To understand how crises turn from possibilities into reality, to paraphrase Marx, we need to move between levels.[7] We need to start somewhere, so let us start at the top.

6 Don Mitchell, 'There's No Such Thing as Culture', p. 109.

7 Karl Marx, *Theories of Surplus-Value*, part 2 (Moscow: Progress Publishers, 1968), pp. 509, 515.

Economic and Ecological Abstractions (Levels 1–3)

The first level of abstraction – the sudden crash, the crisis as shock and event – is the way crises are most often presented in the news. Among economic crises, the first level usually reflects the 'official' names of the crises. The crisis of the 1930s became *the stock market crash of 1929*, the economic crisis of the 1970s became *the oil crisis of 1973*, the crisis that began in 2007 became *the financial crisis of 2008*. The crises appear here as different, created in their own unique historical geographies, preferably related to some external factor. The crises also often have other names, pointing to the next level of abstraction: the designations the *Great Depression* (1930s), the *Great Stagflation* (1970s) and the *Long Depression* (2008–?) show that they are more than bubbles and sudden bangs.[8] When Gérard Duménil and Dominique Lévy, for example, identify four structural crises – the 1890s crisis, the Great Depression, the 1970s crisis and the crisis of neoliberalism – we are on the way to level two.[9] These are more than just events and shocks but crises within hegemonic ways of organising the political economy.

The ecological crises we face are also usually at the first level: storms, wildfires, floods, dead seas and new pandemics. Many of these occurred even before the advent of modern capitalism, but to understand why they now come more frequently, more powerfully and more abruptly – and, in the worst cases, with their own feedback mechanisms – we need to go further into the abstractions.

At the second level of abstraction, we encounter the historical-geographical organisation of the capitalism of which the crises are a part. In other words, this level contains quite a lot: it is about the organisation of work and class struggle, technological developments, energy systems and our historical relationship with nature, circulation and trade, financial and monetary policy, degrees of (and forms of) centralisation in the economy, the creation of new geographical landscapes, and more.

At level two, we encounter the concrete organisation of capitalism. I acknowledge that this level is quite broad and risks being unwieldy,

8 See, e.g., Michael Roberts, *The Long Depression: How It Happened, Why It Happened, and What Happens Next* (Chicago: Haymarket, 2016), pp. 1–8.

9 Gérard Duménil and Dominique Lévy, *The Crisis of Neoliberalism* (Cambridge, MA: Harvard University Press, 2011).

resembling what Don Mitchell calls a 'bad' abstraction. However, I have chosen this approach because it maintains some internal cohesion within the general framework (i.e., the historical-geographical organisation of capitalism) and relatively clear demarcations to the levels above and below. This approach also avoids the mess that's created when there are too many levels of abstraction.

For the ecological crisis, level two is about the concrete organisation of fossil capital, fossil fuel use, global warming, the constant destruction of wilderness and industrialised animal management (creating new zoonoses), more toxic emissions into nature and mass extinctions. If the first level is the ecological crisis as it is presented in the news, this level is perhaps how we encounter the crisis in documentary films or in books. The relationship between the first two levels is central to understanding ecological crises and many people analyse the relationship between these levels, even if often unconsciously.[10] For example, it is a common argument that global warming is leading to more wildfires, although it is difficult to know exactly which fires would not have happened without the 1°C increase in temperature globally (or 1.7°C in Sweden) over the last hundred years. Zoonoses are nothing new, but with capitalism's abuse of nature, they are happening more often. Major storms and floods become more serious events in the context of sea level rise, and crop failures have different consequences when the average temperature rises. Due to development on level two, there are more crises on level one.

When the corona crisis paralysed the world in 2020, many journalists wanted to investigate whether this new pandemic was due to climate change. In other words, will there be more pandemics due to rising average temperatures? What one really wants to know with such a question is whether the effects of the climate crisis (a warmer planet) are among the causes of the corona crisis. They may be, but that is still not the most interesting relationship. If we want to understand why both global warming and pandemics are escalating at the same time, we need to go one step further.

At a third level of abstraction, we find the metabolic rift – that unsustainable relationship between humans and nature. The rift in the metabolism between humans and nature is a prerequisite for the

10 See Armon Rezai and Sigrid Stagl, *Ecological Macroeconomics: Introduction and Review*, Working Paper Series No. 9, 2016, Institute of Ecological Economics, Vienna University of Economics and Business, p. 1.

transformation of nature into capitalist commodities, a premise for all of capitalism's ecological crises. It is precisely the combination of fossil fuels within the metabolic rift of capitalism that has created the gigantic climate crisis we face today. The 2020 corona crisis and the climate crisis must therefore be understood as *two capitalist ecological crises*.

When economic crises shake the world with their creative destruction, it is very much level two that is being revolutionised. This is the level where crises are shaped and become part of capitalism at large, and where economic crises eventually find their solution. It is at this level we see how the crises of the 1930s marked the transition to a Keynesian and Fordist capitalism, and the crises of the 1970s the transition to neoliberalism. Crises that develop and are resolved *within* such periods may rather cement the hegemonic order at level two.

At level two, capitalism can develop in ways that make the economy more or less prone to crisis. There is no reason for different sectors to develop in a coordinated and even manner, which can contribute to destabilising the economy. Disproportionality can arise between different sectors, not infrequently between what is often called the financial sector and the real economy, or between what Marx called department one (production of the means of production) and department two (consumer goods).[11] Over-investment in one sector can also create instability, as in IT up to 2000, or in housing up to – and, let us not forget, even after – 2008. Disproportionality should not be seen as a comprehensive crisis theory in itself, but rather as a general phenomenon that helps us explain differences between sectors and variations across time and space. It is an element of broader uneven and combined development: that capitalism develops differently with different consequences in different places, for different people, etcetera – but always united by underlying trends.[12] At level two, we also find technological development and degrees of competition/centralisation and so on as factors that create more or less instability.

During an economic crisis the spotlight often falls on the financial sector, both as cause (e.g., 1929/2008) and solution (e.g., it must be

11 Karl Marx, *Capital*, vol. 2 (London: Penguin, 1978), chs 20–1.

12 For an introduction to uneven and combined development, see, e.g., Alexander Anievas and Kerem Nişancıoğlu, *How the West Came to Rule: The Geopolitical Origins of Capitalism* (London: Pluto Press, 2015), ch. 2.

reorganised or bailed out to solve the problems). It is tempting to put the whole sector on level two, but it is a bit more complicated than that. The financial sector has a special position in capitalism as a coordinating function between different sectors: it is a sector that can intervene when physical goods cannot become money or when money cannot find places where it can become physical capital; that lends and demands back; that can move money and crises at an extreme pace; that can produce new money; and that often becomes the helping hand of states in crisis management. Money gives social power, and the financial sector is powerful in itself. On the basis of loans and credits, financial innovations and packages, speculation and bubbles, gigantic markets can be created which are potential risks because they must always be backed by *tomorrow's* profits.[13] This sector has therefore a particularly strong position to discipline politicians with the mere threat of disapproving of a political decision.

The financial systems can often – like a 'colossal system of gambling and swindling', in Marx's words – both postpone and amplify future crises.[14] After forty years of neoliberalism, we know the sector can accelerate the development of capitalism, and thus its contradictions and crises. Economic problems can be hidden by financial operations: rotten mortgages can appear valuable when they've been packaged (pre-2008); the economy can be kept going with additional loans (post-2008); and they can produce larger bubbles by creating more money – until the bubble bursts. The post-war era was organised on other principles; a financial sector more under political control can at least to some degree be a stabilising factor in the economy.

The financial sector carries an aura of mystery. It is easy to be seduced by its size and power; it appears complex and difficult to really understand – partly because much financial capital shuns public scrutiny. Because of the central role of the financial sector in capitalism, crises often appear to be financial crises. But why did the financial crashes in 1929 and 2008 trigger such huge economic crises, while the even larger stock market crash on Black Monday in 1987 passed like almost nothing happened? The answer lies in the levels of abstractions.

13 McNally, 'Beyond the False Infinity', pp. 11–12.

14 Karl Marx, *Capital*, vol. 3 (London: Penguin Classics, 1981), p. 572. In relation to ecological macroeconomics, see Rezai and Stagl, *Ecological Macroeconomics*, pp. 7–8. On 'financialisation', see Brett Christophers, 'The Limits to Financialization', *Dialogues in Human Geography* 5, no. 2 (2015), pp. 183–200.

Adam Tooze's *Crashed: How a Decade of Financial Crises Changed the World* has, for many, become *the* book on the 2008 economic crisis. Which is no wonder: it is well written and combines great detail with general analysis and nicely highlights links between politics and finance. Tooze's analysis misses two key aspects: one political and one analytical.[15] Politically, his analysis only relates to the corridors of power. This takes the focus away from the relationship that more than anything defined the years leading up to 2008: how actors with power have dominated and exploited all the rest of us. It then becomes difficult to see how class struggle and social movements can change the world.[16] The analytical weakness is that Tooze does not dive deep enough into the levels of abstractions. In contrast to Marxist approaches, *Crashed* examines very little of what lies beneath the financial sector. How healthy was the underlying economy? Why did American banks throw bad loans at poor people? Why was the financial sector given carte blanche to construct unsustainable systems that accumulated risk? And why did neoliberalism need the bubbles and the crisis to survive?

According to Perry Anderson, Tooze simply 'takes the hypertrophy of finance . . . as a situational given' and the reader is immediately plunged into a sea of events and people where 'structural features emerge only from the point of view of actors attempting to deal with them'. It is as if, Anderson continues, 'decade-by-decade decline in growth of the real economy, across advanced capitalism – the long down-turn that arrived in the seventies – had occurred on another planet'.[17] What Anderson is doing with this critique is forcing us to start thinking about the third level of abstraction. Minor crises and imbalances can be created at level two, but we cannot understand major crises if our analysis stops here. If we want to understand why some economic crises shake some economies for a few days and then disappear, while others send world capitalism into a complete panic and do not seem to let go until the prevailing hegemony is badly shaken, we need to move on to the next level.

Level three I have called 'crisis tendencies inherent in capitalism'. Here,

15 According to Tooze, the book is Keynesian rather than Marxist, cf. the podcast *Politics Theory Other*, 2018, episode 23.

16 Cf. Aditya Chakrabortty, 'The Post-Crash World: How the 2008 Crisis Led to Our Current Age of Extremes', *New Statesman*, 15 August 2018.

17 Perry Anderson, 'Situationism à L'envers?', *New Left Review* 119 (2019), pp. 71–2, 87.

we find tendencies already introduced in the two previous chapters. With the economic crises, we find contradictions that continuously deteriorate the general health of the economy. We made a distinction in chapter 1 between a production perspective (emphasising the increasing organic composition of capital) and a realisation perspective (emphasising overproduction). Both these tendencies come with their counter-tendencies, the biggest of which is the crisis itself – the creative destruction – which shakes up level two and changes the image of capitalism. On level three, in ecological crises, we find the metabolic rift constantly reproducing the premise of ecological crises. What these three underlying tendencies and processes have in common is that they are constitutive elements of capitalism (level three) that makes the concrete organisation of capitalism (level two) erupt in events that we call crises (level one).

I'm not particularly fond of squeezing reality into tables, but when analysis risks slipping into endless complexity, simplification may be necessary to provide clarity. In table 1, I have placed the two crises at five levels of abstraction.

Table 1: Crises of capitalism at five levels of abstraction

Level of abstraction	Economic crisis	Ecological crisis
Level 1 The surface, the event, the shock	'Stock market crash' 1929, 'oil crisis' 1973, 'financial crisis' 2008, 'corona Crisis' 2020.	Heatwaves, storms and forest fires, pandemics, etc.
Level 2 Concrete organisation of capitalism	Historical-geographical organisation of capitalism (and nature). Disproportionality, technological regimes, uneven development.	Historical-geographical organisation of nature (and capital ism). Use of fossil fuels, poisons, capitalist animal husbandry, deforestation.
Level 3 Crisis tendencies inherent in capitalism	Overproduction, the increasing organic composition of capital.	Metabolic rift.
Level 4 Fundamental A	Profit (competition, growth, increased productivity).	
Level 5 Fundamental B	Use-value vs. exchange-value.	

The main relations between economic and ecological crises are *not* as events and shocks. Economic and ecological crises do not cause each other as events or shocks, but this is still how they are very often perceived. So, before coming to the decisive relational connections between the two crises (levels four and five) we need to unpack this argument a bit more.

The Oft-Told Story That Bad Ecology Is Bad Economics

That ecological crises (understood as levels one and two) create economic crises is a fairly common theory. At the risk of being overly speculative, I think a great many socialists, social liberals and social democrats *want* this to be true. And vice versa: they want it to be true that being climate-smart must be economically advantageous. They want the economy to be in good condition when it does not harm nature. Green investment pays; it *has to* pay. What will otherwise happen to our earth? When the humanist-liberal organisation Global Humanitarian Forum argues the world loses $125 billion every year due to climate change, they say implicitly that we would benefit from stopping global warming.[18] Unfortunately, such a figure is only relevant if the world were a static and planned economic entity. In our world, that figure is more likely to conceal the fact that powerful corporations and states are still making huge profits by creating climate change.

The premise of the Brundtland Commission – that ecological problems can only be solved if we also eradicate global poverty – is absolutely relevant in many local places. With global warming, it is rather the other way around: the more poverty, the lower the contribution to climate change. The theory of the environmental Kuznets curve – that richer societies become more environmentally friendly – may lack scientific support, but it *feels* right. After all, money can buy everything, so why not a better environment? Even in the financial sector, there are forces that argue that the fossil fuel industry and climate change will create major economic problems. In April 2019, the governors of the UK's and France's central banks and the head of the Network for Greening the Financial Services (a coalition of thirty-four central banks) warned of the gigantic financial costs caused by climate change. Individual companies that fail to adapt to the new world will lose massively, argued the CEOs, who were particularly concerned that a disorderly transition

18 Global Humanitarian Forum, *The Anatomy of a Silent Crisis* (Geneva: Global Humanitarian Forum, 2009), p. 1.

from fossil fuels to renewables would lead to what they called a 'climate-driven "Minsky moment"' – by which they meant a 'sudden collapse in asset prices'.[19]

That problems related to nature create economic crises is a hypothesis with many variants and a long history. The English economist David Ricardo attributed the cause of the crises in the early nineteenth century to declining agricultural yields and rising grain prices (which led to higher wages and lower profits).[20] For Thomas Malthus, a contemporary of Ricardo's, economic development was limited by the carrying capacity of land.[21] Many years later, the Club of Rome's *Limits to Growth* concluded that future shortages of resources such as energy, minerals and forests, which were either running out or becoming too expensive, would lead to world economic crisis. And, more recently, Jason W. Moore has argued that the fundamental problem of capitalism has been that capital's demand for 'cheap nature' grows faster than its ability to secure it; thus production costs rise and accumulation begins to falter.[22] A crisis then results.

Moore's reasoning is extensive. 'Cheap nature' includes what he calls the cheap four: food, labour, energy and raw materials.[23] To clarify the argument, we need to take the assertions apart and point to examples. For instance, Moore argues that it was the price of energy sources that led to the shift from hydroelectricity to fossil fuels, which historical studies have disproved.[24] High oil prices have historically caused

19 Bank of England, *Open Letter on Climate-Related Financial Risks*. Open letter from Governor of Bank of England Mark Carney, Governor of Banque de France François Villeroy de Galhau and Chair of the Network for Greening the Financial Services Frank Elderson. Bank of England, 17 April 2019.

20 See, e.g., Grossman, 'Marx, Classical Economics', p. 31. Marx had enormous respect for Ricardo, yet claimed that he 'did not actually know anything of crisis', and in a sarcastic critique Marx argued that this 'is the childish babbling of a Say, but it is not worthy of Ricardo' (Marx, *Theories of Surplus-Value*, pp. 497, 502; see also Karl Marx, *Grundrisse* [New York: Vintage, 1973], p. 411).

21 Rezai and Stagl, *Ecological Macroeconomics*, p. 3.

22 See, e.g., Moore, 'Toward a Singular', p. 288; Moore, 'The End of Cheap Nature', p. 285; Jason W. Moore, *Capitalism in the Web of Life: Ecology and the Accumulation of Capital* (London: Verso, 2015).

23 Moore, *Capitalism in the Web*, p. 53. Additionally, Moore even includes 'cheap money' as something that can produce and reproduce cheap nature.

24 See Andreas Malm, *Fossil Capital: The Rise of Steam Power and the Roots of Global Warming* (London: Verso, 2016); Andreas Malm, *The Progress of This Storm: Nature and Society in a Warming World* (London: Verso, 2018), ch. 6.

problems, such as in 1973 and 1979, which supports Moore's hypothesis. In 2020, the problem for the US economy was, rather, that prices were too low as the price of crude oil was in negative territory and companies had to pay to store it. Capitalism seems to be able to survive problems linked to both high and low prices for both oil and labour. Concerning labour, Moore's argument could work as a defence of the profit squeeze theory in the 1970s, but before 2008, excessively low wages were more of a problem for the economy.

Not unlike Moore, writers in the peak oil movement argued that the 2008 crisis erupted because of a reduced supply of oil.[25] The theory of peak oil was inspired in part by *Limits to Growth* and was a hot potato until around 2012. The theory comes in different flavours but the core consists of a handful of assertions: because oil is a finite resource, at some point, global production will peak (which everyone agrees with), which will lead to a deep economic recession in oil-importing countries (which is unclear, but could definitely happen), and this peak has already been reached or will be reached very soon (which is demonstrably wrong). The peak oil movement lost momentum after 2010 when new oil fields opened almost as fast as scientists published texts predicting the end of oil. Unconventional oil production boomed during the 2010s, not least with oil sands and fracking.[26] The fact that oil production expanded radically in the 2010s while the economy was not flourishing supports neither the peak oil movement nor Moore's hypotheses.[27] The environmental activist and author George Monbiot wrote a 2012 column in the *Guardian*, aptly titled, 'We Were Wrong on Peak Oil. There's Enough to Fry Us All.'[28]

The most interesting theorist defending the hypothesis that ecological crises create economic problems was the first-wave eco-Marxist James O'Connor. In 1987, the year before O'Connor launched his theory on the second contradiction, he published a book on economic crises,

25 Moore, *Capitalism in the Web*, p. 1; see, e.g., Richard Douthwaite, 'Degrowth and the Supply of Money in an Energy-Scarce World', *Ecological Economics* 84 (2012), p. 187; Gail E. Tverberg, 'Oil Supply Limits and the Continuing Financial Crisis', *Energy* 37, no. 1 (2012), p. 27.

26 For discussion, see Ugo Bardi, 'Peak Oil, 20 Years Later: Failed Prediction or Useful Insight?', *Energy Research and Social Science* 48 (2019), pp. 257–61.

27 See, e.g., Tverberg, 'Oil Supply Limits', p. 33.

28 George Monbiot, 'We Were Wrong on Peak Oil. There's Enough to Fry Us All', *Guardian*, 2 July 2012.

The Meaning of Crisis: A Theoretical Introduction. This may explain why he turned ecological analysis into a theory of crisis.[29] While economic crises were necessary for capitalism to reproduce itself, the ecological crises were, according to O'Connor, primarily destructive in nature. This is an important conclusion that also underlies this book. O'Connor goes a step further: the destructive ecological crises – global warming, acid rain, salinisation of groundwater, toxic waste, soil erosion, new pesticides, perpetual urban renewal, etcetera – not only destroy people, places and other species, they destroy and undermine profit rates and damage the economy. It would cost so much to repair all the damage to nature – raising the costs of external conditions of production – that it would lead to reduced profits and a crisis for capitalism as a whole. Ecological crises create economic ones.

When I started writing this book during the fall of 2019, my intention was to articulate a severe critique of O'Connor's thesis. Then came the corona crisis. And what was that if not precisely an ecological crisis triggering an economic crisis? Was this the first real O'Connor crisis?[30] At least I needed to be a bit more humble: we need to include in our analysis the fact that ecological crises can be triggers for economic ones. The pandemic did not fully cause the economic crisis (we return to this in chapter 6). It *triggered* it. And, as a main historical tendency, there is still much more that points in the opposite direction: that ecological crises have *not* been problems for capital. It does not matter how much people want 'ecological' to equal 'profitable'. The most important lesson from the history of capitalism is that capital has profited handsomely from creating environmental damage and ecological crises.

Here, too, we must remind ourselves of the flexibility of capitalism. Nature can be destroyed regardless of whether there is a boom – when there is enormous pressure on the earth's resources – or a crisis – when politicians blindly prioritise the economy over everything else in order to restore growth. Capitalism can destroy nature in all sorts of forms, from fascism to social democracy, from neoliberalism to kingships, and

29 James O'Connor, *The Meaning of Crisis: A Theoretical Introduction* (Oxford: Basil Blackwell, 1987); James O'Connor, 'Capitalism, Nature, Socialism: A Theoretical Introduction', *Capitalism Nature Socialism* 1, no. 1 (1988), pp. 15–16; James O'Connor, 'On the Two Contradictions of Capitalism', *Capitalism Nature Socialism* 2, no. 3 (1991), pp. 107–9.

30 See also Andreas Malm, *Corona, Climate, Chronic Emergency: War Communism in the Twenty-First Century* (London: Verso, 2020), p. 112.

under the most diverse political regimes, from Nazi Germany to post-war Sweden, from present-day South Africa to tomorrow's China. Some profit from destroying nature concurrently as others profit from trying to save the planet through insurance, organic products, medicines and psychotropic drugs, waste management, face masks and much more. Capitalism is flexible enough to get out of economic crises, reshape itself and make money from both destroying and saving the planet – all at the same time. One criticism of O'Connor is that capital can accumulate under virtually any ecological conditions, no matter how degraded, at least as long as there are humans still around.[31] Another is that capital can always find its climate change shock doctrine. Where O'Connor argues that it would cost a lot to repair ecological destruction, his critics point out that capital in the world's richest countries always seems to find ways to shift negative consequences to the Global South.[32]

But we cannot ignore the possibility that O'Connor will be vindicated *in the end*. What a world that has warmed up three or four degrees looks like, we cannot know. Will the economic system break down when the last drop of oil is consumed, say in 2120? We will discuss breakdown theories later in this chapter, and leave things here by saying that this is both impossible to predict and politically irrelevant at this point.

One question that is very much alive, however, is whether the capitalist class could continue to accumulate over two degrees of warming. Will capital – if we do not actively stop it – invest all the way into unchecked climate change? Will politicians accept investments that take us into the time of escalating feedback effects? Or perhaps there exist some moral barriers, or at least a self-preservation instinct on the part of capitalists and politicians, that will prevent this from happening? Unfortunately, this is not a hypothetical question. Many investments that will take us into the age of uncontrollable feedback have already been made. In October 2019, Norway's Equinor, formerly Statoil, opened the largest new oil field outside Norway since the 1980s. The field will produce crude oil for fifty

31 Paul Burkett, 'Fusing Red and Green', *Monthly Review*, 1 February 1999; see also Rikard Warlenius, 'Inledning: Fyra debatter och en begravning', in Rikard Warlenius (ed.), *Ecomarxism: Grundtexter* (Stockholm: Tankekraft, 2014), pp. 26–9.

32 Kohei Saito, 'Marx's Theory of Metabolism in the Age of Global Ecological Crisis, Deutscher Prize Memorial Lecture', *Historical Materialism* 28, no. 2 (2020), pp. 16–21; see also John Bellamy Foster, *Marx's Ecology: Materialism and Nature* (New York: NYU Press, 2000), p. 174.

years. From Mozambique to the Arctic, from Qatar and the Kingdom of Saudi Arabia to Canada, oil companies continue to invest as if climate science did not exist. In 2022 and 2023 – as climate change escalated and many had problems paying their energy bills – the big oil companies announced astonishing profit figures.

Nor does it seem difficult to find financial investors. Although the head of the Bank of England seems worried about a 'climate Minsky moment', there is no reason to doubt that financial investors will continue to invest as long as it is profitable, which it still very much is.[33] Listening to the young in the environmental movement, it may seem that the climate issue is a generational conflict, which it also is, as today's youth have inherited a gigantic problem from previous generations. At the same time, there should be no doubt that there are sufficient future capitalists among today's youth who will be ready to invest in the destruction of the planet, well into the age of mass extinction, and enough potential officials and politicians who will support them.

Ecological crisis might have economic consequences, and economic crisis might have ecological consequences. The corona crisis 2020 was an interesting case. However, so far, we have only seen conjunctural relationships between the two crises. Are there also necessary connections? We suggest that there are. Andrew Sayer's concept of diabolical crisis – that when solving one crisis, it becomes harder to solve the other – is interesting. Not only does it point to the practical problems of resolving economic and ecological crises simultaneously, it is also the key to understanding the relationship at a deeper level of abstraction. The relationship is diabolical precisely because both problems are rooted in a system driven by the pursuit of profit. We need, thus, to enter abstraction level four.

Crises as Violent Manifestations of the Contradiction between Use-Value and Exchange-Value (Levels 4–5)

At this level, for the first time in our analysis, the crises are given an identical basis to stand on. The crises acquire a necessary connection.

---

33 Brett Christophers, 'Environmental Beta or How Institutional Investors Think about Climate Change and Fossil Fuel Risk', *Annals of the American Association of Geographers* 109, no. 3 (2019), p. 772.

The eternal pursuit of profit is the main driving force of capital accumulation, and thus of the crises of capitalism. If capital accumulation is the core of capitalism, the crises now go straight to the central nervous system of our political-economic system. Here, we find that the mechanisms increasing productivity, producing growth and apparently providing wealth are the same that produce crises.

It is the eternal quest for increasing profits that drives investment and new rounds of capital accumulation. Zooming in, we see billions of different daily events – capitalists searching for profits in all kinds of ways, workers being exploited by capitalists in different situations, nature being used in a myriad of ways. Zooming out, we see different capitalist activities pulling in the same direction, creating general rates of profit and economic growth at aggregate levels. Capitalists conventionally invest with the intention of producing profits, but growth – in contrast – is produced by actors who normally have no intention of creating growth. On a societal level, the need for both profit and growth is supported by states, politics, ideology and cultural expressions; it is considered a precondition for the reproduction of the system as a whole, and for states in particular.

The capitalist class's perpetual pursuit of profit drives the system towards overproduction and, at the same time, makes companies constantly rationalise and streamline, thus increasing the organic composition of capital. The perpetual pursuit of profit also reproduces and reinforces the metabolic rift.

Now that we are rummaging around in the central nerves of the system, we can even go a step further into the abstractions. At level five, we find the contradiction between exchange-value and use-value. We are now entering a general level where all we can see is what Bensaïd called 'potential crises'. 'Abstract general possibilities' only means abstract crises without content or any compelling motivating factor.[34] A tendency, Mandel reminds us, that does not manifest itself materially and empirically is no tendency at all.[35] To explain a crisis in its most general and abstract form is, according to Marx, 'to explain the crisis by the crisis'.[36]

34 Daniel Bensaïd, 'The Time of Crises (and Cherries)', *Historical Materialism* 24, no. 4 (2016), p. 15; see also Marx, *Theories of Surplus-Value*, p. 509.

35 Ernest Mandel, *Late Capitalism* (London: Verso, 1978), p. 20.

36 Marx, *Theories of Surplus-Value*, p. 502.

We have arrived at the outer perimeter of abstractions. This is not how you explain the recent crisis at a family dinner table, but what we find here is absolutely central to understanding why crises can occur at all.

Anyone with a certain level of social competence understands that you cannot just tell someone who loses their job or their house that the crisis arose because of the contradiction between use- and exchange-value and that it will be resolved when capitalism redefines itself through creative destruction. That is not how you build social movements or launch political alternatives (even though the statement itself is perfectly correct). If crisis theory is to be more than an academic exercise, it needs to be related to ordinary people's issues and problems, and we will return to this later. First, let us dive further into the abstractions.

In *Theories of Surplus-Value*, Marx writes that the most abstract form of crisis is the metamorphosis of the commodity. One expression of this is the contradiction between exchange-value and use-value. Marx brings this out when discussing the theory of overproduction, but we will take it a step further and work from the hypothesis that all crises of capitalism are embedded in the contradiction between exchange-value and use-value.[37]

In their simplest form, use-value and exchange-value are two manifestations of a commodity, or, rather, commodities are bearers of use- and exchange-values.[38] As use-value, the commodity is always concrete and heterogeneous; as exchange-value, it can be highly abstract and homogeneous. In the labour process, we can make a similar division. The use-value of commodities is created by humans using tools, which can be called concrete labour (something humans have been doing, in principle, for thousands of years by using nature to cook and make tools). The specificity of capitalism lies in the exchange-value of commodities: here the concrete and specific disappears. The key point is that one commodity must be equated with another commodity. For this purpose, money is used, but money (or the form of money) cannot give us the answer to why one copy of commodity X can be exchanged for two commodities Y. In Marxism, the central point is that what gives commodities their exchange-value is that they reflect the labour that is

37 Ibid., p. 509.
38 Harvey, *A Companion*, p. 16.

put into the commodity – the labour value. This corresponds not to the actual hours of labour (in which case, a good would, by definition, be more expensive the longer it took to produce it) but to the general socially necessary labour time in the production process. This is called abstract labour.[39] On the one hand, we have the concrete labour that actually produces things, on the other hand the abstract labour that creates surplus value and defines capitalist production.

The concepts of use-value and exchange-value can also be used to describe other general phenomena. For example, we can say that capitalist production is based on what gives profit (exchange-value) and not on what people need (use-value). Health care, urban planning, housing and much else are increasingly organised according to the principle of exchange-value and less and less according to use-value. The French sociologist and philosopher Henri Lefebvre discussed how cities are increasingly organised according to the imperative of capital accumulation (exchange-value), at the expense of all of us who use cities for our daily lives (use-value). Even more generally, he showed how capitalism created global abstract spaces for the production, circulation and distribution of capital.[40] With global banking and financial systems, containerisation, a global aviation industry and much more, capital can operate (almost) anywhere on Earth and still feel at home. When both time and space are organised according to how they can benefit capital accumulation, both abstract space and abstract time are created, in stark contrast to human lives which are still about concrete places (space) and biological ageing (time). Harvey discusses the conceptual pair in relation to Marx's theory of value: where use-value can be described in terms of absolute time and space, exchange-value can be understood in terms of the relational time and space of the world market.[41]

The creation of abstract time and space would have been impossible if money did not exist as a universal equivalent. (Writing about money

39 Marx, *Capital*, vol. 1, ch. 1; see, e.g., Harvey, *A Companion*, ch. 1; Guglielmo Carchedi, *Behind the Crisis: Marx's Dialectics of Value and Knowledge* (Chicago: Haymarket, 2012). For a discussion of use-value and exchange-value in relation to Marxist crisis theory, see also Grossman, 'Marx, Classical Economics', pp. 6–83; Kuhn, 'Henryk Grossman', pp. 74–6.

40 Lefebvre, *The Production*, ch. 6; see also Neil Smith, *Uneven Development: Nature, Capital, and the Production of Space* (Oxford: Basil Blackwell, 1991), pp. 111–26.

41 Harvey, *A Companion*, p. 37.

is difficult, as it is 'the god among commodities' and 'the permanent commodity', in Marx's words, in a form that can seduce us all and has also become a means of power in itself. It may be tempting to think that capitalism is most profoundly about money, which it is not.[42]) The tension between labour value and its form of representation – money – comes itself as an unstable relation. Through economic policies or financial bubbles, for example, a lot of money can be created, but it cannot grow forever because it must relate to the basic building block of economic growth over time: increasing labour value.

When commodities go to markets, they all meet as the money form. As a universal equivalent, money is thus an externalisation and a product of the inherent contradiction between use-value and exchange-value of the commodity. The very dividing of the commodity itself into commodity and money creates a general possibility of capitalist crises.[43]

The financial system, in which money is apparently allowed to flow freely independently of labour or nature, may appear to be the purest form of capitalism: a 'capitalist dream-world in which capital infinitely produces itself out of itself'.[44] Here, money can seemingly multiply without use-values – without the ugly industry, the grumpy worker, and almost entirely without nature. Without any obstacles or barriers, capital seems to have found a way into infinity.

Contradictions between use-value and exchange-value can be kept in check in the short term by the appearance of money as a third party, a bridge between the particular and the universal. Rather than a 'mediating institution', we can say with Marx and Hegel that money represents a 'pseudo-mediation', since it is a material form of this contradiction.[45] Money, like politics, can postpone contradictions – but always at the risk of reinforcing the problems.

In the process of turning capital into more capital, the interest in the use-value is lost. The important thing is that capital must be accumulated so that even more capital can be accumulated, no matter what is accumulated. From the perspective of the capitalist class, accumulation

42 Marx, *Grundrisse*, pp. 221, 231.

43 Ernest Mandel, *The Second Slump: A Marxist Analysis of Recession in the Seventies* (London: NLB, 1978), p. 167; see also Harvey, *A Companion*, p. 36.

44 McNally, 'Beyond the False Infinity', p. 12.

45 Ibid., pp. 16–18.

takes on an abstract and quantitative character. Accumulation for the sake of accumulation, production for the sake of production.[46] Capital accumulation is a process without any end, and with money as the universal equivalent the aim is to get as much money as possible. No upper limit can be defined. There is no ceiling. The more the better is the rule to live by, because competitors do the same. The purpose and meaning of accumulating capital are non-articulated goals, never-articulated goals, or in Marx's words: abstract wealth.[47]

We began this chapter by discussing abstraction as a method; now we have also encountered abstraction in two other forms. One is concrete or real abstractions, that is, non-physical but highly real phenomena with concrete expression. The last is abstraction as a utopian idea; as a description of a process seemingly without material anchorage; as a project for shaping the world that (attempts to) disregard concrete and physical conditions. These are three different forms of abstractions, but all point to one of the origins of the verb 'abstract', namely to separate. As a method, we must separate to create clarity; as a concrete abstraction, value is separated from use-value; and as utopian idea, the daily and earthy experiences for most of us are separated from the aim of the ruling class: abstract wealth.

The three forms of abstraction have different relations to material reality from which they can never escape. When we use abstraction as a method, the whole point is to try to capture reality – to which degree an abstraction is good or not depends on whether it captures the material world. Concrete or real abstractions can never be understood as completely divorced from material conditions: when we talk about exchange-value, we are always implying actual people and nature that produced it.[48]

Capitalism builds its world on utopian ideals of infinite growth and organises itself and our world according to principles of exchange-value, while operating in a highly concrete world. Most commodities are still physical: labour is concrete human bodies, and nature is as tangible as ever. No ideology or philosophical thought process can change this, and failure to understand the difficult relationship with the concrete will

46 Marx, *Capital*, vol. 1, p. 742.
47 Marx, *Theories of Surplus-Value*, p. 503; Marx, *Capital*, vol. 1, p. 253.
48 See Grossman, 'Marx, Classical Economics', p. 28.

lead to empty or unreal abstractions.[49] Use-value is a constant reminder that we are, after all, human beings living on a small planet. Capitalism is founded on the contradiction of having infinity as a horizon within a materially limited reality. This is seeking *abstract prosperity* in a *concrete world.*

Capital accumulation is driven towards infinity – it tries to make itself into what David McNally calls an 'absolute abstraction' – but can never escape the fact that it is also use-value: value can only expand over time as both use-value and exchange-value.[50] Capital accumulation carries with it the promise of the infinite – because there is no final destination – but this is a *false infinity*. Through new rounds of capital accumulation, capitalism is driven towards an infinity it can never find, precisely because the system depends on a material and physical base. This false infinity clashes with the non-infinite, creating both crises and misery for people and nature.[51] Perhaps the greatest contradiction of all in capitalism is that it is created by people on a specific geographical planet, but driven by a ruling class that must pretend that concrete people and physical nature do not exist.

We cannot escape the fact that use-value is, in one way or another, a constraint on capital.[52] Intuitively, it seems that this must be a problem for capitalism. Surely a constraint must be considered a concern? Here we have to turn the coin around. Rather than a problem, the clashes between use-values and exchange-values and the crises these produce are *prerequisites* for capitalism.

If we take the contradiction between use-value and exchange-value and return to the crises at abstraction level three, we see that the contradiction also binds the levels together. Overproduction is an overproduction of exchange-values, not of things people necessarily need. This is not just about the (in)ability to consume all goods produced, but, rather, about consuming all goods at a price that yields a satisfactory profit. From the perspective of many poor people, we should, rather, be talking about the underproduction of basic use-values such as food, medicine,

49 Marx, *Grundrisse*, pp. 81–114; Grossman, 'Marx, Classical Economics', p. 28.

50 McNally, 'Beyond the False Infinity', pp. 8, 10.

51 For a good discussion in relation to crises, see ibid., pp. 1–23.

52 On introducing *scarcity* into Marxist crisis theory without becoming Malthusian, see O'Connor, 'Capitalism, Nature', p. 26.

housing. From an ecological perspective, the suffering of the world is just as much about over-consumption. Overproduction, from the perspective of the capitalist system, is only possible because production is organised precisely according to exchange-value. Had it been organised by use-values, or if consumers had direct access to use-values, the general tendencies towards overproduction could not occur.

As the organic composition of capital increases, the economy will be more prone to crisis. The key point here is that labour value depends on a distinction between exchange-value (related to abstract labour) and use-value (concrete labour). With rationalisations, increased productivity and new technology, the capitalist class is keen to get rid of concrete labour. Capital accumulation would be easier without workers, these disturbing, unreliable and untrustworthy elements who can even go on strikes and engage in class struggle from below. Capitalists want to reap the fruits from abstract labour, while excluding concrete labour. When the actual and concrete worker is removed, abstract labour will also disappear, which means less surplus value and an unhealthier economy – another reminder that we can never ignore the material dimension.

The contradiction between use-value and exchange-value is probably most striking in the context of the ecological crisis. Oil, gas, coal, fish, animals, land and green forests are all commodities – exchange-values – that can be bought on the stock market, and also part of ecosystems in which nature reproduces itself without asking the capitalists for advice. That nature has use-value for humans is nothing new: we have always used nature within a social metabolism.[53] That nature has also become exchange-value for humans – that it has been incorporated into the economy as capital, as raw materials, commodities, investment objects and so on – that came with capitalism. This is the essence of the metabolic rift. If nature had been merely a stock (an exchange-value) on the stock exchange, the ecological footprint would be limited; then we would not have had any metabolism between humans and nature at all. If nature had only been a use-value, something that could not be bought and sold, there would never have been a rift in the social metabolism.

The never-ending chase and desire for profits and endless growth will

53 On use-value as transhistorical concept, see Carl Cassegård, *Toward a Critical Theory of Nature: Capital, Ecology, and Dialectics* (London: Bloomsbury Publishing, 2021), ch. 5.

always be in a dialectical tension with a reality that materialises in physical commodities, labour and nature. It is often said that ecological crises come from a system based on limitless capital accumulation on a finite planet. This is true. Now we can also add that this is even true concerning economic crises.

Use-value and exchange-value are two phenomena that have become seemingly independent of each other, but, in every crisis, they are forcibly reunited; they are drawn towards each other and collide. Capitalist crises are recurrent examples of the fact that exchange-value cannot exist without use-value. The crises are rooted in the fact that eternal and endless accumulation always meets a reality it can never escape. Use-value and exchange-value must be understood as simultaneously separated and united. One cannot divide a physical commodity into two parts and say that one part is use-value and the other exchange-value, or divide the working day into two and say that the first half is concrete work and the second half abstract. This is also the case with use-value and exchange-value in crises. If they were only united without being separated, no violent division and basis for the crises would have been possible. If they were merely separate without being united, they would never meet and thus not cause crisis. We can therefore say about the crisis of capitalism what Marx argued about circulation and production: they are the 'forcible establishment of unity between elements that have become independent and the enforced separation on one another of elements which are essentially one'.[54] We have an economy pulling away, heading for infinity, but repeatedly encountering a material reality. The crises are the response of the concrete to the cry of the abstract; or the response of death to a system in which the rulers have begun to believe in eternal life.

To Shake and to Stabilise

When everything seems to be pregnant with its contrary, we must analyse capitalism as dialectically as it actually is. To understand how and how much crises affect society is no easy task. Because crises are extraordinary events that go down in the history books and processes that dramatically change the world, it can be tempting to attribute too

54 Marx, *Theories of Surplus-Value*, p. 513.

much importance to them. Political regimes, technological development, cultural expressions, revolutions, geopolitical situations and much else can be read into crises. This is not necessarily wrong, but it is often difficult to know how accurate such interpretations really are. Crises are indeed periods of great change, but there are also a lot of changes in the absence of a crisis. A great deal does *not* change during crises. Certain things even remain unchanged because of crisis, when things have to change to stay the way they are.[55] One aspect of capitalism that never changes is that it always has to change.

Change and continuity in capitalism can be discussed with our five levels of abstraction. The general trend is that, while crises shake the world at level one and deep crises transform level two, there are fewer changes at level three, as the processes and tendencies here are permanent in capitalism. Levels four and five consist of fundamental elements of capitalism that are stable factors in crisis after crisis.

Crises have a double function; they both shake capitalism and keep the system alive. Crises both change and consolidate. Capitalism needs the turbulence to survive. Here we must immediately make a distinction between our two crises. In economic crises, the crisis itself is part of the solution to the crisis. Capitalism needs creative destruction to restore rates of profit, shake things up and reinvent itself. Ecological crises are different. Capitalism does not need the ecological crises as events (level one). It would function just as well without wildfires, sea level rise or global warming. But it must *produce* the ecological crises.

When discussing how capitalism is reproduced through change, it is interesting to look at differences between the post-war period and neoliberalism in the Global North. At first glance, it is tempting to conclude that the post-war period was a better form of capitalism than neoliberalism: overall growth was about 3 per cent per capita in the post-war period, compared to 1.4 per cent per year after the 1980s; productivity increases were much higher during the 'golden age of capitalism'; and, according to the IMF's definition of recession, there were no recessions between 1945 and 1973 compared to six between 1973

55 For reflections on change and non-change under capitalism, see, e.g., Daniel Bensaïd, *Marx for Our Times: Adventures and Misadventures of a Critique* (London: Verso, 2002).

and 2010.[56] While the post-war period is known as the golden age, neoliberalism has staggered from one crisis to the next. However, if neoliberalism has been an inferior version of capitalism, why has it existed *longer* than the cherished Fordist-Keynesian era? One answer concerns class power, which we will revisit in the next chapter. Another is that the concept of a golden age is US- and Euro-centric, as it ignores how other economies, like China's, can be crucial for the reproduction of various forms of capitalism. A third answer, that we will discuss further here, is that neoliberalism survived longer *thanks to the crises.*

The crises of the 1980s and 1990s in Asia, Latin America and Russia – and the whole neoliberal policy of the IMF and World Bank in Africa during the same period – occurred under such strong hegemony that they never shook the imperialist core countries. The crises were used by Washington, the IMF and the World Bank to further consolidate neoliberal hegemony. This was the time of 'structural adjustment', when problems were solved with new loans which came with demands for further financial liberalisations, deregulations and privatisations.[57] This tendency must be read geographically: the crises in Asia, Russia and Latin America affected countries that – in various ways and in various degrees – were more or less dependent on imperialism's headquarters in Washington. Even in the Global North, neoliberalism was reproduced through bubbles and crises – from the Volcker shock to the financial and banking crises of the 1980s and 1990s, to the IT crash of 2000.[58] While the crises between the 1980s and 2006 were used to legitimise and expand the neoliberal regime, the 2008 crisis carried a seed with the potential to burst the entire neoliberal hegemony.

If neoliberalism was relatively robust thanks to the crises, this puts the debate about neoliberalism as an instable system into new perspective. Because there were lower profit rates in many sectors, weaker economic growth and more crises during neoliberalism than the

56 See also, e.g., Harvey, *The Enigma*, p. 8; Andrew Sayer, *Why We Can't Afford the Rich* (Bristol: Policy Press, 2015), p. 193.

57 See, e.g., Naomi Klein, *The Shock Doctrine: The Rise of Disaster Capitalism* (London: Penguin, 2007), ch. 8; Paul Burkett and Martin Hart-Landsberg, 'Crisis and Recovery in East Asia: The Limits of Capitalist Development', *Historical Materialism* 8, no. 1 (2001), pp. 9, 31.

58 See also Robert Brenner, 'What's Good for Goldman Sachs Is Good for America: The Origins of the Current Crisis', Institute for Social Science Research working paper, UCLA, 2009.

post-war period, neoliberalism has been characterised as a 'constant crisis'.[59] Perhaps we should rather formulate a different hypothesis, which may seem self-contradictory, but is not: because neoliberalism was more crisis-prone, it was more robust as a hegemonic system. Due to its crises, it was *not* in permanent crisis.

From Crisis to Collapse?

The main arguments in this book are quite different from those that see in the crises the key to the great collapse. Kautsky saw overproduction not only as a recurring phenomenon, but also as a secular tendency. He believed that instability would increase continuously until it reached the ultimate point, resulting in the collapse of capitalism. (The political task, from this perspective, became to build a movement capable of taking power on the day when capitalism made itself redundant. Simon Clarke stresses that Kautsky's critique of the Russian Revolution must be seen in this light; the waiting for the 'decisive moment' led to both political passivity and bureaucratic degeneration within the workers' movement.)[60] The downfall of the system was certain, according to Kautsky: the 'moment is drawing near when the markets of the industrial countries can no longer be extended and will begin to contract'. This 'would mean the bankruptcy of the whole capitalist system'.[61] Kautsky wrote this in 1910.

Writers who see the collapse of capitalism in a crystal ball often combine different approaches. The first approach, and perhaps the most important, is to emphasise a trend or a (crisis) tendency and then take it to its logical conclusion. But this ignores an ocean of context. To put it

59 For good discussion, see David McNally, *Global Slump: The Economics and Politics of Crisis and Resistance* (Oakland: PM Press, 2011), ch. 2.

60 Simon Clarke, *Marx's Theory of Crisis* (London: St Martin's, 1994).

61 Karl Kautsky, *The Class Struggle (Erfurt Program)* (Chicago: C. H. Kerr, 1910), pp. 84, 87; see also Rosa Luxemburg, 'Our Program and the Political Situation', in Peter Hudis and Kevin B. Anderson (eds), *The Rosa Luxemburg Reader* (London: Monthly Review Press, 2004), pp. 357–72; Rosa Luxemburg, *The Complete Works of Rosa Luxemburg*, vol. 2, *Economic Writings 2*, ed. Peter Hudis and Paul Le Blanc (London: Verso, 2015), p. 302; see also Kuhn, 'Henryk Grossman', pp. 61–3; Bill Dunn, 'Marxist Crisis Theory and the Need to Explain Both Sides of Capitalism's Cyclicity', *Rethinking Marxism* 23, no. 4 (2011), p. 532; Norman Geras, *The Legacy of Rosa Luxemburg* (London: Verso, 2015), ch. 1.

bluntly, for Paul Mason it is the cost of production: if marginal cost is (close to) zero – that is, when one more unit of a given good can be replicated almost free of cost – how can capital make profits? It seems impossible, so surely capitalism must fall? Jason W. Moore points out that capitalism is reproduced because of the 'cheap four': and, if nature is no longer 'cheap', and if this is a precondition for capitalism, then it follows that capitalism breaks down. The second way of moving from crisis to collapse is to emphasise why and how there will be crises, while ignoring how crises are solved and managed. A third approach is to anoint the whole thing in a general belief that everything that is bad will get worse (in all honesty, a not-unknown phenomenon among left intellectuals).

There are different paths to the collapse, but a fascinating common aspect of each theory is that the collapse is predicted to come during the author's lifetime. A brilliant exception here is Henryk Grossman, who is notorious for his mathematical proof that capitalism would fall after thirty-four (!) cycles of 5 to 10 years, i.e., between 170 and 340 years in the future. (In Grossmann's defence, this was a thought experiment done as a critique of Otto Bauer's scheme which, just as idiotically, would 'prove' by mathematical calculation that capitalism could continue forever.)[62] Most people, however, tend to foresee the collapse in their own time. Moore is clearly right that the price of food, labour, energy and raw materials is central to the development of capitalism. Rather than flexible and dynamic processes, this is, for Moore, a linear and determinate progression that has *just now* culminated in the 'breakdown of the strategies and relations that have sustained capital accumulation over the past five centuries.'[63]

According to Swedish economic historian Rasmus Fleischer, the idea that capitalism digs its own grave all the way to collapse has been taboo among almost all twentieth-century Marxists. According to Fleischer, both social democrats and communists perceived these crisis theories as a threat because 'breakdown theories seemed to leave no room for a historical subject. If capitalism automatically conducts its own demise,

62 On how Grossman considered capitalism's tendency to break down to manifest in the form of recurring crises rather than an uninterrupted collapse, and how he did not propose a theory of the 'automatic collapse of capitalism', see Kuhn, 'Henryk Grossman', pp. 70–89. See also Mandel, *Late Capitalism*, p. 31.

63 Moore, *Capitalism in the Web*, p. 1.

what would be the task of the labour movement?'[64] The last question is certainly pertinent. But a taboo? The main problem for breakdown theory is, rather, that everyone who ever plugged away at it has been proven wrong. A breakdown theory that ignores the role of historical subjects – either working class or capitalist class – is unpopular not because of taboos, but rather because it is really a blind alley.

But this book is not about the possible breakdown of capitalism, so I will not go further into breakdown theories here. Breakdowns can also come from tendencies and processes that are not directly related to how I define crisis, so I will not go as deep into, for example, Moore's and Fleischer's theories as they perhaps deserve. That must be left for elsewhere.

At the opposite pole of collapse theory, we find David Harvey: 'Capitalism will *never* fall on its own. It will have to be pushed.'[65] Maybe. But who knows? We do know that capitalism will, one day, be replaced by another mode of production, and people will then need theories to explain that. But no one can know today what this transition will look like, hence we should also remain humble about the future. Still, I contend that contemporary theories about the fall of capitalism are primarily guesswork. If someone manages to hit the nail on the head, I think this has less to do with their genius, and more to do with luck.

For these purposes, I will only conclude that there is not much in the economic crises that seems to drive the capitalist system towards any collapse. Based on our discussions on creative destruction and crises as contradictions between convulsive and stabilising forces, I think it is a far more plausible hypothesis that economic crises *reproduce* capitalism. But what about the ecological ones? They tend to escalate, do they not?

Climate change changes everything, they say. Could global warming prove Kautsky right after all? Right that *some* internal contradiction would bring down capitalism? Again – and perhaps we might add, sadly – the answer seems to be no. Here, we need to be clear about what we are talking about. Global warming may indeed contribute to the collapse of many human societies, but would it lead to the collapse of capitalism as a system? If we – for the sake of the argument – say that the world explodes due to climate change, capitalism will obviously disappear too.

64 Rasmus Fleischer, 'Värdekritisk kristeori: att tänka kapitalets sammanbrott', *Fronesis* 46–7 (2014), p. 90, our translation.

65 Harvey, *The Enigma*, p. 260, emphasis added.

But the relevant question is whether capitalism will collapse before this. Climate change will bring massive upheaval to many sectors that are crucial for the reproduction of capitalism including food production, energy, transport, health and so on. Will these changes be so massive that they will bring down the whole system? Again, we cannot know. But doubt is justified. Considering the 'enormous elasticity of capital', Kohei Saito argued, 'it remains unclear whether capitalism or the Earth will collapse first'.[66] Perhaps it is more likely, to put it bluntly, that the last capitalist will sell a jug of gasoline to his last customer in a world on fire; or that the last capitalist will order workers to use the latest technology to produce even more survival kits. But I might be wrong. However, if ecological crises – hypothetically – were to undermine capitalism in a chaotic and apocalyptic process, it would happen so far into the future that it is politically irrelevant for any reader of this book.

It is more urgent to change perspective and ask: Will capitalism *live longer* because of climate change? Can the climate crisis serve as some kind of creative destruction for the overall economy? The climate crisis has some components that make these questions relevant. The climate crisis destroys nature, but from the perspective of capitalism, it is capital (value) that must be devalued. And the climate crisis does that, too. A study from 2016 argued, for example, that $2.5 trillion in financial assets could be erased due to extreme weather, temperature increases, drought and other consequences of climate change. In a worst-case scenario, 17 per cent of all the world's assets could be lost, which the researchers behind the report say would devastate the global economy.[67] We will not go into the maths here; our question is whether such massive devaluations of capital tend to break the capitalist system. The history of the economic crises suggests quite the opposite.

Capitalism thrives on change, and the climate crisis will open up gigantic opportunities for capital accumulation. New markets, business opportunities, geographies and technologies will emerge; poverty may keep wages down; new consumption patterns will be needed, and so on. A whole new (global) infrastructure will have to be built with railways,

---

66 Kohei Saito, *Marx in the Anthropocene: Towards the Idea of Degrowth Communism* (Cambridge: Cambridge University Press, 2022), p. 127.

67 See Damian Carrington, 'Climate Change Will Wipe $2.5tn Off Global Financial Assets', *Guardian*, 4 April 2016.

ports and energy systems. Capitalists can first accumulate capital when producing the crisis, and then further when trying to stop it. When the world has given up, there will be opportunities for privatised *geoengineering* (for example, large-scale manipulation of the earth's climate system by dimming the sun's rays or changing the composition of gases in the atmosphere). This bastardised form of creative destruction is a bit different from the one discussed in context of the economic crisis. Where economic creative destruction saves the economy, the 'ecological' version does not save the ecology. It also saves the economy.

After all this, it is worth repeating that we should be humble when discussing the future. We do not know exactly what the next crisis will look like; certainly not capitalism's doomsday. The best starting point, both analytically and politically, is to assume that capitalism can exist under any ecological regime; it will just work differently. No one can know how/if capitalism can be reproduced with temperature increases of four degrees. But there are at least as many good arguments for capitalism to exist in such a warm world as there are for any other mode of production.

The Road to Crisis Is Paved with Rational Decisions

When climate change escalates or millions become unemployed in a matter of weeks, the word 'irrationality' comes to mind. It is hard to imagine how this could not be due to extreme systemic failures. That is not the case. The crises are the result of many actors at different times doing exactly what they thought was rational. If capitalists do what capitalists should do, if workers do what workers should do, if we treat nature according to all existing laws and rules, the system will continue to run into crisis after crisis.

After more than 200 years of recurring economic crises, the obvious question from a liberal perspective is: Can they really be *so* difficult to avoid? Under laissez-faire, politicians tried to wait out the crises; people inspired by Keynes have tried to regulate away the crises; and under neoliberalism, many tried to privatise them away. We might even add that fascism has tried to bomb the crises away.[68] No one has succeeded. Over and over again, we run into old patterns of booms, overinvestment, crashes and panics.

68 See reference to Gramsci in chapter 5.

Mandel aptly asked in the 1970s whether the booms and busts derive from 'some irrational "herd instinct"' within the capitalist class.[69] The answer he gave is as simple as it is prescient: what is rational from the perspective of individual large firms is irrational from the perspective of the system as a whole. Mandel pointed to a similar contradiction in his analyses of ecology: when production is for profit – abstract exchange-value – it can be just as rational for individual firms to produce junk or weapons as food and medicine; there is rationality within the various parts of the economy and irrationality for the socio-economic whole.[70]

Other theorists have pointed to similar phenomena. According to Georg Lukács, the irrationality of capitalist production as a whole is qualitatively different from the laws that regulate various parts of the economy. In Herbert Marcuse's 'one-dimensional' society, 'its sweeping rationality, which propels efficiency and growth, is itself irrational'. David Harvey argues that crises bring with them an 'irrational rationality' and, according to Theodor Adorno, society is gaining more and more control over its inhabitants, but this is happening in parallel with a growing irrationality: 'The world is not only mad, it is mad and rational.' Enzo Traverso draws on the concepts of rationality from Weber, Lukács and Mandel when he considers Auschwitz from this point of view: as a deadly example of a combination of partial rationality (*Teilrationalität*) and total irrationality (*Gesamtirrationalität*) that he considers typical of advanced capitalism.[71] Capitalist crises are monumental collisions of different forms of rationality.

Economic crises are created by actors who, even if they would theoretically accept the whole analysis in this book, cannot avoid the crisis themselves. In capitalism, the ruling class is doomed to create crises. In economic crises, we can see how *partial rationality* clashes with *total irrationality* at different levels of abstraction. In the case of overproduction, companies have to face their problems through further efficiency

69 Mandel, *The Second Slump*, p. 178.

70 Mandel, *Late Capitalism*, p. 508.

71 Georg Lukács, *History and Class Consciousness: Studies in Marxist Dialectics* (Pontypool: Merlin Press, 1971), p. 102; Herbert Marcuse, *One-Dimensional Man: Studies in the Ideology of Advanced Industrial Society* (Boston: Beacon Press, 1964), p. xiii; David Harvey, *Seventeen Contradictions and the End of Capitalism* (London: Profile Books, 2014), p. ix; Theodor Adorno and Max Horkheimer, *Towards a New Manifesto* (London: Verso, 2011), pp. 38–40; Enzo Traverso, *Understanding the Nazi Genocide: Marxism after Auschwitz* (London: Pluto Press, 1999), p. 55.

gains, by producing more and cheaper while lowering wages. In order to survive, they have to take measures that reinforce the general problem. The organic composition of capital increases because all firms *must* increase efficiency, rationalise and find new technology to keep up with the competition, but the system as a whole gets into trouble when everyone implements the new technology. We also see the rationality/irrationality contradiction at abstraction level two, with disproportionality between sectors of capitalism. Since all companies have to increase profits, industrial companies have also invested more and more in financial markets under neoliberalism. This, too, is rational for each individual industrial firm, as profits were higher here, but created ever greater imbalances and instability in the economy as a whole.

Even on the surface of crises, we can locate similar phenomena, which, in turn, is why the phenomenon appears in liberal textbooks. Here, rationality/irrationality is often understood through what is called the Minsky moment, named after the American economist Hyman Minsky.[72] In short, the theory is that, when the economy is doing well, firms have to invest more and more because they are driven by competition. Even less risk-prone firms have to keep up and contribute to increased risk taking because their competitors have already done so. The result is growing bubbles and over-investment. When the behaviour and thus the bubbles peak, it will be difficult to sell the most speculative assets. When the tide turns, companies will be forced to sell even the less speculative assets, which, in turn, will only reinforce a general downward trend in the market. Then everyone has to sell as much and as quickly as possible. The bigger the bubble, the bigger the blow. This is why it is often said that there are two contradictory rules in economic crises: 'do not panic' and 'panic before everyone else'. In context of

72 For (critical) analyses, see Jan Kregel, *Minsky's Cushions of Safety: Systemic Risk and the Crisis in the US Subprime Mortgage Market* (Levy Economics Institute, Economics Public Policy Brief Archive, no. 93, January 2008); Gary A. Dymski, 'Why the Subprime Crisis Is Different: A Minskyian Approach', *Cambridge Journal of Economics* 34 (2010), pp. 239–55; Timur Behlul, 'Was It Really a Minsky Moment?', *Journal of Post Keynesian Economics* 34, no. 1 (2011), pp. 137–58; Riccardo Bellofiore, 'Crisis Theory and the Great Recession: A Personal Journey, from Marx to Minsky', in Paul Zarembka and Radhika Desai (eds), *Revitalizing Marxist Theory for Today's Capitalism* (Leeds: Emerald Group Publishing Limited, 2011), pp. 81–120; Charles P. Kindleberger, *Manias, Panics, and Crashes: A History of Financial Crises* (New York: John Wiley and Sons, Inc., 2000), ch. 2.

capitalism at large, some capitalists should not invest so much when things go up and should not sell when the bubbles burst. Capitalists who chose that path soon became bankrupt.

The psychology of the crisis is also an important component for many Keynesians. If bad news appears on the horizon – it could be increased debt or rumours that a boom is a bubble – people become less confident in the economy and spend less. Uncertainty spreads and investors shy away from investing; the prediction becomes truth, a self-fulfilling prophecy, and thus a crisis.[73]

Concepts such as 'mass psychosis', 'delirium' and 'hysteria' go back at least to the Tulip Bubble in Amsterdam in the 1630s and the bubbles in France and England in the eighteenth century.[74] As Galbraith so lyrically points out, there are always lots of people who think that they are destined to get rich without having to work, that they have some special luck or divine advantage, special access to inside information or perhaps exceptional financial acumen.[75] The 'Minsky moment' is – in my view – a sophisticated way of understanding irrational bubbles and crashes within a rational framework. Personal psychology plays a role, but it must always be understood within the framework of capitalism. And speculation, as Štajner reminds us, is 'the inevitable outcome of capitalist accumulation'.[76]

The most dramatic expression of the clash between the rational and the irrational, Mandel wrote in the 1970s, was the organisation of humanity's potential collective suicide through nuclear war and disaster

73 Cf. Derek Wall, *Economics after Capitalism: A Guide to the Ruins and Road to the Future* (London: Pluto, 2015), p. 22; see also John Maynard Keynes, *The General Theory of Employment, Interest and Money* (London: First Harvest, 1964); John Maynard Keynes, *Essays in Persuasion* (New York: Classic House Books, 2009). On Keynesians seeing the main cause of economic crises in the realm of theories, ideas and thoughts, see Robert Skidelsky, *Keynes: The Return of a Master* (New York: Public Affairs, 2009), pp. xiv, 169. For Schumpeterian critique, see Chris Freeman and Francisco Louçã, *As Time Goes By: From the Industrial Revolutions to the Information Revolution* (Oxford: Oxford University Press, 2001), ch. 2. For Marxist critique, see Paul Mattick Sr, *Marx and Keynes: The Limits of the Mixed Economy* (London: Merlin, 1971).

74 Lars Magnusson, *Finanskrascher: från kapitalismens födelse till Lehman Brothers* (Stockholm: Natur & Kultur, 2020), p. 66.

75 John Kenneth Galbraith, *The Great Crash 1929* (London: Penguin, 2009), p. 13.

76 Rikard Štajner, *Crisis: Anatomy of Contemporary Crises and (a) Theory of Crises in the Neo-imperialist Stage of Capitalism* (Belgrade: KOMUNIST, 1976), p. 46; see also Kindleberger, *Manias, Panics, and Crashes*, ch. 2.

according to the most advanced economic principles.[77] Today, the climate crisis is the ultimate expression: rational decisions have succeeded each other, then suddenly we find ourselves on the verge of a new geological epoch. Environmental crime comes in all sorts of forms, but the main rule is that the creation of climate change has been a perfectly legal and well-organised process. It is administered and implemented by well-trained people. Fossil capital with its global infrastructure is, arguably, one of the most well-organised and well-functioning parts of what we call our civilisation. What clash, according to Daniel Bensaïd, are the immanent rationality of biology and the mechanical reason of the market.[78] What is rational for each company is, on balance, a disaster for the environment.

Politics follows a different logic from economics, but the clash is similar. Since capitalist countries need to secure economic growth for their states to function, it becomes rational for each individual country to reduce emissions as little as possible. Ironic – or, rather, tragic – examples are Iran or Louisiana, USA. Both have short-term interests in increased oil production, but Iran will suffer terribly from global warming and more and more of Louisiana will soon be underwater.

Seemingly rational crisis management can also go wrong, and the levels of abstraction can help us understand why. In the midst of a fire, firefighting feels undeniably rational. What actually constitutes 'rational' crisis management is a difficult question, not least because no one knows exactly how a crisis will unfold. With the climate crisis, cities and countries are prepared to build higher levees to protect against floods, or to deploy the military against wildfires, but even rich countries affected by global warming do little or nothing about the underlying causes of the crisis. In the face of economic crises, banks and powerful companies/sectors are often rescued in an attempt to stop the crises from escalating. During the economic crisis of the 1970s, Sweden was quick with counter-cyclical policies, and after the 2008 crisis, most states were extremely quick to rescue their financial systems. But similar to the way increased subsidies to industrial production in Sweden in the 1980s could only postpone the fall of

77 See Mandel, *Late Capitalism*, p. 510.

78 Michael Löwy, 'Daniel Bensaïd: A Marxism of Bifurcation', *International Viewpoint*, 28 June 2020.

Fordism, ever larger subsidies to banks or lower repo rates could not save neoliberalism during the 2010s.

When we discuss rationality here, we must remember that we are talking about a short-term capitalist rationality. Exploiting others, seeking maximum profit and destroying nature are not a universal maxim for rational human action as such. We have reached a point, concluded Baran and Sweezy back in 1966, and this is just as true today, 'where the only true rationality lies in action to overthrow what has become a hopelessly irrational system'.[79]

With crises seemingly all around us, and ecological crises certainly escalating, it is tempting to ask whether capitalism has become even more destructive and ruinous. Has the barbaric side of capitalism become more prominent as it expands and ages? Has the system gone from creative destruction to destructive creativity, or to creative (self-) destruction? Have we reached the point in the development of capitalism that Marx and Engels pointed to in *The German Ideology* of 1846, where the productive forces are no longer primarily productive, but destructive?[80]

In a world on fire, it is certainly tempting to say yes. But even previous progress and advancement of capitalist civilisation has served up atrocities, cruelties and crimes in the past: colonialism, imperialism, concentration camps. Rather than whether more or less productive or destructive, we must ask: For whom? Capitalism's division between progress and destruction has always had a class character. Climate change is indeed destructive for an incredible number of people, but it is happening at the same time as the rich are getting richer. While climate change kills the poor, the system becomes more productive for others. Ecological crises, like colonialism and imperialism before, can present opportunities for the world's rulers. This is not the first time that mass murder has proved to be a good business idea. We need to talk about class.

---

79 Paul Baran and Paul Sweezy, *Monopoly Capital: An Essay on the American Economic and Social Order* (New York: Monthly Review Press, 1966), p. 349, see also pp. 324–5.

80 Karl Marx and Friedrich Engels, *The German Ideology* (Guilford: Prometheus, 1998), p. 60; see also Bensaïd, *Marx for Our Times*, p. 32; Angus, *Facing the Anthropocene*, pp. 108, 126; Marx, *Capital*, vol. 1, p. 638.

4

The Class Character of Crises

Two questions have fascinated me throughout the years I have been working on crises. First, why do they emerge? Second, if crises are man-made and have deadly consequences, why are none of those responsible ever punished? These seem to be two separate questions; the first question is about social analysis, the second about law. The closer we look, the more we see that they are in fact closely linked. The answer to the first question is in the second, and the answer to the second is in the first. And both answers are about class.

Many people never want to talk about class. When there is no crisis, they talk about the 'rising tide lifting all boats'. When there is a crisis, we are told that we are all 'in the same boat'. When there is no crisis, we are told that everyone benefits from growth and profit because money, resources and entrepreneurial spirit trickle down to everyone. When there is a crisis, we must stand united in these difficult times. We rarely hear that the crises are caused by a specific class. The crises are *nobody's* fault, as economic crises occur due to external events, psychology or chance. Alternatively, the crises are *everyone's* fault, as when everyone who has ever consumed anything at all is held responsible for the climate crisis. Since crises are created within an actually existing class society, played out within a class society and resolved within a class society, it would be extremely strange if crises *did not* have a class character. No matter how much people refuse to talk about the class nature of the crises, it still persists. No discourse in the world can dispel this fact.

But what do we mean when we talk about class? If we just take a quick look at history, we see that class is constantly changing. The left waited in vain for the old rhythm of class struggle to return, wrote Stuart Hall in the 1980s, but the left missed the fact that the very forms of class struggle itself were changing.[1] For all the movement, change and complexity, we will nevertheless start here from a simple main distinction, or rather main relationship: between those who own the means of production and those who do not. The former are the capitalist class and the latter the working class. We will complicate and nuance this, but the starting point for analysis is always important. And this is our starting point.

But if class society is constantly changing, is it not rather dogmatic to hold on to such an old Marxist starting point? Archaic though it may sound, the question of ownership brings something quite central to the analysis. For, while much is in flux, not everything is changing. Again, we need to analyse the problem at different levels of abstraction. Class relations can be revolutionised on the surface while underlying relations remain the same; class society both changes and does not change. According to Bensaïd, the ever-changing forms of social differentiation only make it *all the more* necessary to understand the fundamental relations between classes.[2] If we want to grasp a class society that is constantly changing, it is particularly important to try to understand even what is *not* changing. New crises affect new people in new ways, but underneath all the crises is a common thread: the power of the capitalist class over both workers and nature.

Whereas most liberal theories of class focus on inequalities in wealth and income as the cause and preferably the very definition of class, here we will see inequalities rather as a consequence of class society.[3] Even liberals must accept the empirical truth that the rich are richer than the poor, but we must go further and claim that some are rich precisely because others are not. To explain this, Marxists often use the keywords 'exploitation' (workers being paid less than the exchange-value they

1 Stuart Hall, 'Gramsci and Us', in *The Hard Road to Renewal: Thatcherism and the Crisis of the Left* (1988), published on versobooks.com, 10 February 2017.

2 Daniel Bensaïd, *Marx for Our Times: Adventures and Misadventures of a Critique* (London: Verso, 2002), p. 173.

3 Ibid.; Göran Therborn, *Kapitalet, överheten och alla vi andra* (Lund: Arkiv förlag, 2018), p. 25.

have produced) and 'domination' (the ability to control the activities of others).[4] For Marx, capital is the power to command labour and its products: 'The capitalist possesses this power not on account of his personal or human properties but in so far as he is an *owner of* capital.'[5]

Similar to the way we analysed crises previously, we will also start here from Marx without getting bogged down in discussions about what Marx 'really' meant.[6] We will begin with the question of ownership, as this is the best starting point if we want to understand fundamental relations between classes including 1) class's central function in capitalism; 2) the way class can change on the surface while being cemented at the underlying level; and 3) who creates the crises of capitalism. Yet the question of ownership certainly cannot provide all the answers we seek. When it comes to issues such as inequalities between and indeed *within* classes, we should also include other inputs inspired in part by Max Weber (emphasising inclusion in and exclusion from markets), and also pure stratification analyses (e.g., that *x* has more money than *y*). We will return to all this later in the chapter, to blurred boundaries *between* classes and much more. First, let us dig a little into class and crisis.

4 Guillaume Durou argues Marxists have theorised exploitation to a greater extent and with better precision than domination. See Guillaume Durou, 'Class Is What Capitalism Makes of It: Challenging the Lure of "Realism" in Mainstream Class Analysis', *Critical Sociology* 46, no. 2 (2020), p. 218; see also Erik Olin Wright, *Understanding Class* (London: Verso, 2015), p. 9; Michael Löwy, *Fire Alarm: Reading Walter Benjamin's 'On the Concept of History'* (London: Verso, 2005), p. 75.

5 Karl Marx, *Economic and Philosophic Manuscripts of 1844*, in *Early Writings* (London: Penguin, 1975), p. 295. For discussions on class in relation to ownership, power and consciousness, see Stephen Resnick and Richard Wolff, 'The Diversity of Class Analyses: A Critique of Erik Olin Wright and Beyond', *Critical Sociology* 29, no. 1 (2003), pp. 11–13; Wright, *Understanding Class*, p. 189. On ownership and property for Weber, see Max Weber, *Economy and Society: An Outline of Interpretive Sociology* (Berkeley: University of California Press, 1978), p. 927.

6 Stephen Resnick and Richard Wolff aptly commented that Marx, like 'most other sophisticated social thinkers', has used 'different singular concepts of class at various points in [his] work and/or combined them into various composites'. Resnick and Wolff, 'The Diversity of Class Analyses', p. 11.

Losing, Winning and Causing Crises

North American environmentalist Bill McKibben realised early on that climate change is first and foremost killing those who have done the least to cause it.[7] Today, this is undeniable. In what follows, we will see that exactly the same pattern is also found in economic crises. Those with the least power to shape and make decisions about the economy are hit hardest, time after time. The ruling class would probably have been more willing to solve the climate crisis if its members had been systematically killed first – not non-white and poor people as is the case in reality. It may be a morbid thought, but the question is not unreasonable because the killing comes with an impressive precision that targets skin colour and class. How would the surviving big capitalists have reacted if the first to die from ecological crises had been the richest capitalists, the biggest property owners, those with the largest stock portfolios, the most highly paid consultants and the highest executives? We shall leave the question unanswered because, as we all know, it is completely counterfactual.

That economic and ecological crises are man-made and that we live in a class society are two relatively uncontroversial claims. Yet it seems difficult to put two and two together. In what follows, we will do just that by discussing the class nature of crises from three different perspectives: who loses, who gains, and who causes the crises. Moreover, the fact that the crises turn out to have a very similar class character gives us an additional perspective on how these are precisely two *capitalist* crises. In what follows, I hope to convince the reader that, despite all the differences and complexities within and between the crises, the general hypothesis holds: the crises of capitalism hit those who have done the least to cause them hardest, and vice versa.

Losing the Crisis: Poor Die First

But don't even the rich take quite a beating during a crisis? It is certainly true that rich individuals can lose a larger amount of money compared to those with less wealth, because they have more to lose. One of the

7 See Andreas Malm, *Corona, Climate, Chronic Emergency: War Communism in the Twenty-First Century* (London: Verso, 2020), p. 20.

biggest losers after the 2008 crisis, ABC News was able to report, was Maurice 'Hank' Greenberg. He was CEO of the financial and insurance company AIG for twenty-seven years and lost 95 per cent of his private assets during the crisis.[8] Economic crises really do send shock waves through society's wealthier strata: contracts signed are cancelled, companies go bankrupt, business leaders lose jobs and huge investments go up in smoke in seconds.

One narrative that often recurs in economic crises is how rich people lose everything and then choose to take their own lives. A famous example is Sweden's own Ivar Kreuger, who (presumably) killed himself because of the economic problems of the 1930s crisis. In March 2020, there was much media attention when Thomas Schäfer, the fifty-four-year-old finance minister of the German state of Hesse, committed suicide, probably because of concerns about Covid-19 and the economic crash. Perhaps the most famous story about the Great Depression is that of a series of speculators in New York committing suicide by throwing themselves out of windows, but, according to Galbraith, this is a myth. It probably never happened. Nevertheless, these suicides have become the very epitome of the Great Depression.[9] The many workers who actually commit suicide during economic depressions, however, rarely make it into the headlines.

Suicide, Engels wrote in his description of the English working class in 1844, was formerly 'the enviable privilege of the upper class', but, in the growing industrial cities, it has also 'become fashionable among the English workers, and numbers of the poor kill themselves to avoid the misery from which they see no other means of escape'.[10] There is clear evidence in the research that increased unemployment during economic crises leads to more suicides.[11] And even the climate crisis is taking

8 Scott Mayerowitz, '2008's Financial Winners and Losers', ABC News, 23 December 2008.

9 On the suicide myth, see John Kenneth Galbraith, *The Great Crash 1929* (London: Penguin, 2009), ch. 8.

10 Friedrich Engels, *The Condition of the Working Class in England in 1844* (Mansfield: Martino Publishing, 2013), p. 116.

11 Tamma A. Carleton, 'Crop-Damaging Temperatures Increase Suicide Rates in India', *Proceedings of the National Academy of Sciences* 114, no. 33 (2017), pp. 8746–51. On unemployment protection and fewer suicides during crises, see Thor Norström and Hans Grönqvist, 'The Great Recession, Unemployment and Suicide', *Journal of Epidemiology and Community Health* 69, no. 2 (2015), pp. 110–16; Rikard Štajner, *Crisis:*

lives: environmental economist Tamma Carleton estimates that 60,000 farmers in India alone have committed suicide in the last thirty years due to higher temperatures. As we move from the capitalist class to the working class or poor farmers, we simultaneously move from exceptional stories to mere statistics.

Rich people *can* lose in crises. But we must also look at how quickly one may recover. The super-rich needed an average of three years to recover from the 2008 crisis, while many workers never came out of the crisis at all and were in a very precarious position when it returned in 2020.[12] 'Hank' Greenberg lost perhaps 95 per cent of his fortune, but ABC News also reported that he was still sitting on hundreds of millions of dollars, with offices on Park Avenue, houses in both New York City and Brewster, NY, and a private jet. Hank's case is extreme, but not unique. Even those who are a little less wealthy usually have something to fall back on, in stark contrast to a poor person who can lose a job and a place to live in a situation where it's already hard to pay for food and clothes for the children.

The broad working class suffers in a qualitatively different way from the rich. It is within the working class that we find the real losers in the crises: those who are hurled out into unemployment and made homeless, who lose *all* their savings and their *entire* pensions (if they had any). Children of workers and farmers in poor countries drop out of school and go without clean water and food. It is the world's poorest who are forced to move, either to the nearest shanty town or across deserts or seas.

During the Asian crisis of the 1990s, 24 million people lost their jobs, according to the International Labour Organization (ILO).[13] For some, this was not really a crisis at all, but rather a blessing. According to the *Economist*, 'it took a national crisis for South Korea to turn from an inward-looking nation to one that embraced foreign capital, change and competition'.[14] All the human misery and the sad fates of the 24 million

Anatomy of Contemporary Crises and (a) Theory of Crises in the Neo-imperialist Stage of Capitalism (Belgrade: KOMUNIST, 1976), p. 81.

12 Pål Velo, 'Formuene eser ut i krisa', *Klassekampen*, 6 May 2020, pp. 14–15.

13 ILO, 'Governing Body to Examine Response to Asia Crisis', International Labour Organization, press release, 16 March 1999.

14 Quoted in Naomi Klein, *The Shock Doctrine: The Rise of Disaster Capitalism* (London: Penguin, 2007), p. 278.

people and their families apparently did not concern the editorial writers of the *Economist* very deeply. And, to be honest, a figure like 24 million losing their jobs is hard to take in, it passes easily by. It is hard to envision, even in the most vivid imagination, 24 million people and their families in a precarious situation. Something that's not as easily ignored – at least not for people with normal emotional functions – is when Naomi Klein describes the social consequences that followed the crises in Asia: mothers sending their children to look for something edible among the garbage; child prostitution in Thailand increasing by 20 per cent the year after the IMF reforms; all those rural families in the Philippines and South Korea who were forced to sell their daughters to international human traffickers who put them to work in the sex industries of Australia, Europe and North America.[15]

In relatively rich countries with social safety nets, the most extreme consequences are avoided, but even in a country like Sweden, it is within the working class that we find those who lose the most. In the *Economic and Philosophic Manuscripts*, Marx writes that the worker does not necessarily win when the capitalist wins, but always loses when the capitalist loses: 'No one suffers so *cruelly* in its decline as the working class.'[16] Perhaps the main common concern for the working class during economic crises is the increase in unemployment. North American economists Carmen Reinhart and Kenneth Rogoff show that, during fourteen banking crises between 1929 and 2001, unemployment rose by an average of 7 per cent in the five years following the onset of the crises.[17] In 1933, 31 per cent were unemployed in the US, and almost 45 per cent in a Germany that had just recovered from the inflation of the 1920s.[18] Because of the 2008 crisis, somewhere between 27 and 40 million people lost their jobs globally. It is an astounding number, yet it pales in comparison to 2020 when 38 million people lost their jobs

15 Ibid., pp. 272–3.

16 Marx, *Economic and Philosophic Manuscripts*, pp. 283–4.

17 The crises they point to are Argentina, 2001; Colombia, 1998; Malaysia, Indonesia, Thailand, Philippines, Hong Kong, Korea, all 1997; Japan, 1992; Sweden, 1991; Norway, 1987; Spain, 1977; Finland, 1991; and the US, 1929. It is worth noting that ten of the crises were in the 1990s, and the authors acknowledge difficulties in measuring unemployment in countries with a large informal sector. Carmen Reinhart and Kenneth Rogoff, *This Time Is Different: Eight Centuries of Financial Folly* (Princeton, NJ: Princeton University Press, 2009), pp. 229, 405.

18 Štajner, *Crisis*, p. 79.

within nine weeks only in the US.[19] With crises comes great danger of increased poverty, and rising food prices in crises make the situation even worse.[20] If you belong to the working class, you also risk losing your home. More than 9 million families lost their homes in the US after the 2008 crisis.[21] The poorest, again, lost the most, and it is impossible to ignore the fact that the crisis took place in a racist society. It is not at all surprising that the non-white poor in the US lost the most. While the median wealth of the Hispanic population, according to Adam Tooze, fell by no less than 86.3 per cent from 2007 to 2010, virtually all the housing wealth of the African American population was wiped out.[22]

We recognise the pattern everywhere; the most precarious fall hardest. Mandel wrote back in 1978 that the 1970s crisis hit women, racialised people and migrant workers harder than others.[23] Women, despite large geographical differences, typically have lower incomes, are more often uneducated and are more often outside the labour market and politics than men, making them more vulnerable in times of crisis. Globally, as many as 70 per cent of the 1.3 billion people living in extreme poverty (less than a dollar a day) are women, and women own only 1 per cent of the world's property. This makes them more vulnerable to both economic and ecological crises.[24] Class societies are racialised, and immigrant workers or undocumented workers are typically

19 For 2008, see ILO, *Global Employment Trends 2011: The Challenge of a Jobs Recovery* (Geneva: International Labour Office, 2011). For 2020, see Patricia Cohen, 'Many Jobs May Vanish Forever as Layoffs Mount', *New York Times*, 11 June 2020.

20 For discussions on poverty, health and food prices during the '90s crisis in Asia, see, e.g., M. Ramesh, 'Economic Crisis and Its Social Impacts', *Global Social Policy* 9, no. 1 (2009), pp. 80–2; Paul Burkett and Martin Hart-Landsberg, 'Crisis and Recovery in East Asia: The Limits of Capitalist Development', *Historical Materialism* 8, no. 1 (2001), p. 30.

21 Adam Tooze, *Crashed: How a Decade of Financial Crises Changed the World* (New York: Viking, 2018), pp. 156–7.

22 Ibid., pp. 47, 156–7. On class and racism in the US housing crash of 2007–08, see Manuel B. Aalbers (ed.), *Subprime Cities: The Political Economy of Mortgage Markets* (Malden, MA: Wiley-Blackwell, 2012).

23 Ernest Mandel, *The Second Slump: A Marxist Analysis of Recession in the Seventies* (London: NLB, 1978), pp. 89–90.

24 Rhona MacDonald, 'How Women Were Affected by the Tsunami: A Perspective from Oxfam', *PLOS Medicine* 2, no. 6 (2005), e178; Senay Habtezion, *Overview of Linkages between Gender and Climate Change*, United Nations Development Programme, Gender and Climate Change Policy Brief (2012). For the Asian crisis of the 1990s, see Burkett and Hart-Landsberg, 'Crisis and Recovery', p. 31.

the first to lose their jobs in the event of redundancy; they are the ones most often left without a contract or safety net in the first place. The capitalist class in countries like Sweden can live with many workers being organised, but it is always imperative to have a lower layer of the working class – a reserve army – that can be drawn in and out as the economy fluctuates. These are very often racialised, immigrant and undocumented workers who can be used – and sacrificed – by the capitalist class during crises.[25]

The Swedish Riksbanken (federal bank) offers an example of the conventional way of measuring losses in crises. They calculated that without the financial crises in Sweden in the 1990s and 2008, GDP would have been SEK 1,000 billion higher than today – about SEK 100,000 per Swede.[26] This is, of course, interesting because money really does disappear in crises. Apart from the fact that the lost sum would clearly not be distributed equally to every citizen, this calculation also ignores the important role of crises in transforming capitalism. A counter-argument to this kind of calculation is that, *without* such crises, there would be no capitalism at all.

While economic crises can have deadly consequences for workers and the poor, ecological crises take killing to another level. Here, too, rich individuals can be affected, but they are still the exception that proves the rule. Climate change has been called a death sentence for millions of people in Africa, Asia and Latin America, but the poor dying from ecological crises is nothing new.[27] We can trace these types of deaths back to the rapid capitalist urbanisation of the nineteenth century that created extreme poverty and disease. Many people know that English cities often have a richer west end and a poorer east end; fewer know

25 See, e.g., Ernest Mandel, 'Introduction', in Karl Marx, *Capital*, vol. 2 (London: Penguin, 1978), p. 22.

26 See Andreas Cervenka, *Vad gör en bank?* (Stockholm: Natur & Kultur, 2017), p. 12.

27 Ian Angus, *Facing the Anthropocene: Fossil Capitalism and the Crisis of the Earth System* (New York: Monthly Review Press, 2016), p. 96. On ecological crises hitting the poor hardest see, e.g., James O'Connor, 'Capitalism, Nature, Socialism: A Theoretical Introduction', *Capitalism Nature Socialism* 1, no. 1 (1988), p. 37; Joel Wainwright and Geoff Mann, *Climate Leviathan: A Political Theory of Our Planetary Future* (London: Verso, 2020), p. 73; Michael D. Yates, *Can the Working Class Change the World?* (New York: Monthly Review Press, 2018), p. 55.

that this is because factories were often located in the middle of the city centre and the wind from the Atlantic blew the pollution to the eastern parts of the cities. This allowed the geography of the city to determine who would live long and prosper, and who would die prematurely.

Natural disasters strike directly into existing class societies. In fact, there are no 'natural' disasters, since a hurricane that destroys nature without any significance for mankind is rarely described as a natural disaster. Phil O'Keefe and two colleagues concluded in *Nature* back in 1976 that natural disasters occur at the very intersection of extreme physical phenomena and vulnerable people. What kills most people in earthquakes is not primarily the earthquake itself, but the collapse of the buildings where people live and work, or the lack of clean water, food and medicine. In Mexico City, houses built on rock are priced higher than those on more unstable ground. After the 1976 Guatemala earthquake, which killed 23,000 people and left 1.5 million homeless, social inequalities became so prominent that local survivors began calling the event a *classquake*.[28]

Heatwaves, extreme weather, floods, poor harvests or other consequences of climate change – they all kill poor and working people before they affect the rich. And women are more vulnerable than men, non-whites more than whites.[29] Many indigenous peoples have livelihood systems and lifestyles that are extremely vulnerable to climate change, not least floods and droughts.[30] The poorest have the least ability to

28 Phil O'Keefe, Ken Westgate and Ben Wisner, 'Taking the Naturalness Out of Natural Disasters', *Nature* 260 (1976), p. 556; see also Neil Smith, 'There's No Such Thing as a Natural Disaster', items.ssrc.org, 11 June 2006. For analyses of the 2010 Haiti classquake, see Charlie Kimber, 'Hell in Haiti as Aid Turns to Occupation', *Socialist Worker*, 26 January 2010. For Nepal 2015, see Andrew Nelson, 'Classquake: What the Global Media Missed in Nepal Earthquake Coverage', *Conversation*, 4 May 2015. On earthquakes and climate change, see, e.g., Göran Ekström, Meredith Nettles and Victor C. Tsai, 'Seasonality and Increasing Frequency of Greenland Glacial Earthquakes', *Science* 311, no. 5768 (2006), pp. 1756–8; Bill McGuire, 'How Climate Change Triggers Earthquakes, Tsunamis and Volcanoes', *Guardian*, 16 October 2016.

29 For a broader discussion, see Petra Tschakert, 'The Role of Inequality in Climate-Poverty Debates', World Bank: Policy Research Working Paper, No. 7677 (2016), pp. 2–3.

30 Laura Pulido, 'Racism and the Anthropocene', in Gregg Mitman, Marco Armiero and Robert S. Emmett (eds), *Future Remains: A Cabinet of Curiosities for the Anthropocene* (Chicago: University of Chicago Press, 2020), pp. 119–20; see also Global Humanitarian Forum, *The Anatomy of a Silent Crisis* (Geneva: Global Humanitarian Forum, 2009), pp. 1, 3.

move, to protect themselves from extreme heat, to have a home office during a pandemic, and they have underlying diseases that make them more vulnerable to other problems, and so on and so on and so on.

The human death rate from climate change is extremely tricky to calculate. World Health Organization argued in 2018 that climate change was expected to cause an estimated 250,000 deaths per year between 2030 and 2050 due to malnutrition, malaria, diarrhoea and heat stress. However, epidemiologists and health researchers such as Andy Haines and Kristie Ebi have criticised that figure for being overly conservative, as it does not include deaths caused by other climate-sensitive health effects, people being forced to flee, lower agricultural productivity or the effects of disruption to health services caused by extreme weather and climate events.[31] The professors Joshua M. Pearce and Richard Parncutt have argued that if global warming reaches or exceeds two degrees, 'mainly richer humans will be responsible for killing roughly 1 billion mainly poorer humans through anthropogenic global warming, which is comparable with involuntary or negligent manslaughter'.[32] Millions will also die from *producing* climate change. A study published in 2021 examined outdoor fine-particle pollution caused by burning fossil fuels. The result was a staggering 8.7 million deaths in 2018, a greater global death toll than tobacco smoking and malaria combined.[33] We can never know exactly how many people will die from climate change, but what we do know is that the current economic policies are pushing us straight towards mass suffering, and that this is unevenly distributed, both between and within countries.

The class nature of crises can be summed up in three words: crises strike downwards. Mandel's words on the class character of economic crisis are still highly relevant, and can even be applied on ecological crisis:

31 WHO, 'Climate Change and Health', who.int, 1 February 2018; Andy Haines and Kristie Ebi, 'The Imperative for Climate Action to Protect Health', *New England Journal of Medicine* 380, no. 3 (2019), p. 266; Rachael Rettner, 'More Than 250,000 People May Die Each Year Due to Climate Change', livescience.com, 17 January 2019.

32 Joshua M. Pearce and Richard Parncutt, 'Quantifying Global Greenhouse Gas Emissions in Human Deaths to Guide Energy Policy', *Energies* 16, no. 16 (2023), 6074.

33 See Karn Vohra et al., 'Global Mortality from Outdoor Fine Particle Pollution Generated by Fossil Fuel Combustion: Results from GEOS-Chem', *Environmental Research* 195 (2021), 110754; Oliver Milman, '"Invisible Killer": Fossil Fuels Caused 8.7m Deaths Globally in 2018, Research Finds', *Guardian*, 9 February 2021.

> [Any crisis] strikes the weak more harshly than the strong, the poor more harshly than the rich. This is true within the imperialist countries themselves, as the crisis affects the proletariat and the bourgeoisie. It is true within the employer class vis-à-vis the small and middle-sized companies on the one hand and the great monopolies on the other. And it is true on a world scale, vis-à-vis the semi-colonial and dependent countries on the one hand and the imperialist countries on the other.[34]

The ecological crises of our time are placed as an additional burden on poor people and countries. First, the Global South was subjected to colonialism, imperialism and robbing of natural resources, then, for some decades, to 'free markets', and now it is workers and small farmers in the poorest countries who are paying the most for climate change. The implications of this have not sunk in everywhere. 'Unlike the crises of capitalism,' wrote historian Dipesh Chakrabarty in 2009, 'there are no lifeboats [in the climate crisis] for the rich and privileged.'[35] This may be true 400 years from now. Or maybe not. What is relevant is that every time the climate crisis shows its ugly face – through wildfires, storms, droughts and so on – in the decades when global warming must be stopped, there *are* lifeboats for the richest.

Winning the Crisis: King of the Crisis

In the spring of 2002, I was lying on a beach in Brazil after volunteering with the Landless Peasant Workers' Movement (MST) for six months, and struck up a conversation with an Argentinean man strolling by wearing a gold watch and a solid-gold piece of jewellery. I could not help but ask about the great debt crisis that was raging in Argentina at the time. There was no problem, he told me, as he had just taken his money and moved to Brazil. When the euro crisis hit Greece in 2010 – a crisis that, among other things, caused a malaria epidemic for the first time

34 Mandel, *The Second Slump*, p. 46.

35 Dipesh Chakrabarty, 'The Climate of History: Four Theses', *Critical Inquiry* 35 (2009), p. 221. For criticism, see Andreas Malm and Alf Hornborg, 'The Geology of Mankind? A Critique of the Anthropocene Narrative', *Anthropocene Review* 1, no. 1 (2014), p. 66; Françoise Vergès, 'Racial Capitalocene', in Gaye Theresa Johnson and Alex Lubin (eds), *Futures of Black Radicalism* (London: Verso, 2017), pp. 74–6.

since the 1970s; saw Europe hit by an HIV epidemic for the first time in a long time, right in the centre of Athens; and led to a dramatic rise in the number of injecting drug users – we could read in the *Guardian* that house prices were rising in some parts of London. While banks in Greece put limits on how much money ordinary people could withdraw in cash, estate agents in London reported that the city's upmarket areas were overflowing with wealthy Greeks looking to put their savings into property. Marx's claim that every crisis temporarily reduces luxury consumption has not exactly held true for recent crises.[36] There are indeed ways to *avoid* crises, but there are also ways to *win* them.

Let us first see if this also applies to workers and the poor – can they 'win' and be better off because of the crises of capitalism? As individuals and as exceptional cases, yes. Petra Tschakert, a professor of rural development, concludes in a report for the World Bank that there is very little evidence of the positive effects of climate change on poor people. However, she adds, there may be some exceptions, things she calls 'isolated events', such as the accumulation of social assets, agricultural diversification, and collective action. Instead, Tschakert presents strong evidence that the wealthy *often* exploit shocks and crises with their flexible assets and power status.[37]

But let us not drop the question entirely, because it also deserves to be asked on a more general level: Can workers also benefit from the destruction of nature? Are not growth and profit also the basis for workers' wages, pensions and welfare? This is true, to a certain extent. When capital grows and destroys nature, workers can benefit. This does not change the fact that the working class gains the least when the economy is doing well, and loses the most in crises.

In urban history, it is often stressed that the sanitation and health problems that followed early urbanisation in England – with London as a case in point – were the beginning of urban reform, planning and social programmes. This is not wrong, but this history is about much more than helping the poor. Investment in toilets and sewers only began when it was realised that water-borne diseases couldn't be avoided in the rich areas unless the poor also had sanitation. When uncontrolled

36 See Helena Smith, 'Greek Wealth Finds a Home in London', *Guardian*, 13 April 2010; Marx, *Capital*, vol. 2, p. 486.

37 Tschakert, 'The Role of Inequality', p. 42.

urbanisation appeared to be creating a 'dangerous class' that was difficult to manage, the importance of organised urban planning became clear. One main argument for using planning to prevent fires from spreading was to keep property prices up in the nicer parts of town. At the same time, healthy workers were needed in the factories and in the military. The ruling class simply could not live with a working class that was 'biologically unfit' for social reproduction.[38] Thus arose urban planning and public health, partly to safeguard the health of the workers, but primarily to safeguard the profits of the rich and the nation.

After up to 10,000 people died in Hamburg in less than two months during the cholera epidemic of 1892 – equivalent to 13 per cent of the city's population, most of them clearly poor and working-class – the left went on to win the next election. This led to better housing and cleaner water systems being built.[39] From this, of course, one can conclude that the poor were better off thanks to the cholera epidemic. Like they had been after the Black Death, the survivors were better off because the supply of labour was reduced. (Even though a study of fifteen different pandemics in Europe, from the Black Death [1347–52] to the swine flu [H1N1 pandemic] in 2009, shows that only moderate wage increases have taken place after such pandemics.)[40] But if the only way the poor and working class can 'win' an ecological crisis is by having their friends and family die, then I think most will feel the price is too high. This is not an equation the political left should accept. This is the barbarism of capitalism exemplified. If we are to find the real winners of the crises, we need to focus on those who win without paying such a price. Say, for example, John Paulson, the hedge fund manager and investor who made $15 billion in 2007 and a few billion more the following year simply by betting that the US mortgage market would collapse.[41]

38 Peter G. Hall, *Cities of Tomorrow: An Intellectual History of Urban Planning and Design in the Twentieth Century* (Oxford: Blackwell, 1988); Peter Marcuse, 'Housing in Early City Planning', *Journal of Urban History* 6, no. 2 (1980), pp. 153–76.

39 Olga Khazan, 'How the Coronavirus Could Create a New Working Class', *Atlantic*, 15 April 2020.

40 Òscar Jordà, Sanjay R. Singh and Alan M. Taylor, *Longer-Run Economic Consequences of Pandemics*, Working Paper No. 26934, National Bureau of Economic Research, 2020, p. 1.

41 Gregory Zuckerman, *Greatest Trade Ever: The Behind-the-Scenes Story of How John Paulson Defied Wall Street and Made Financial History* (New York: Crown Business, 2009).

Speculators can also profit from creating crises by shorting a currency. The most famous example of this is George Soros on Black Wednesday, 16 September 1992. Soros's Quantum Fund initiated a massive gamble on the morning that the UK could no longer defend the exchange rate and needed to devalue, and thus leave the European Exchange Rate Mechanism. Like a self-fulfilling prophecy, Soros cracked the pound, made a billion US dollars, depressed the British pound and humiliated British politicians.[42] And who had to pay? It is estimated that Black Wednesday cost British taxpayers $3.3 billion. Similar speculation was also central to the creation of the Asian crises five years later. This time, Goldman Sachs, J.P. Morgan and Soros were all involved. The most important player now was GE Capital, the financial arm of General Electric, which, in the *Far Eastern Economic Review*, simply went by the name 'King of the Crisis'.[43]

One company that has profited massively from creating ecological crises is Exxon. The company is notorious for having known for decades that their emissions are changing the climate, but lying to politicians, hiding the truth and producing alternative facts, all to avoid focusing on the issue, and all while making huge profits.[44] Anyone who has made good money from investing in fossil fuels and/or their infrastructure is a winner in the climate crisis. And, while mortality rates increase and human suffering surges, we can surely say that this cycle is far from over.

One strategy for winning both economic and ecological crises is to use the shock and panic that accompanies crises to push through self-interest, change societies and cement power. In *The Shock Doctrine*, author Naomi Klein presents some particularly stark observations on how those in power under neoliberalism have cynically used and created crises to usurp even more power over people and nature. Although the basic idea is not entirely new – Niccolò Machiavelli, for

42 For an overview, see Rohin Dhar, 'The Trade of the Century: When George Soros Broke the British Pound', priceonomics.com, 17 June 2016.

43 Burkett and Hart-Landsberg, 'Crisis and Recovery', p. 26.

44 See exxonknew.org; Andreas Malm and the Zetkin Collective, *White Skin, Black Fuel: On the Danger of Fossil Fascism* (London: Verso, 2021), ch. 1; Naomi Oreskes and Erik M. Conway, *Merchants of Doubt: How a Handful of Scientists Obscured the Truth on Issues from Tobacco Smoke to Global Warming* (New York: Bloomsbury Press, 2010).

example, argued that 'atrocities must be carried out at once' – *The Shock Doctrine* shows how it became almost institutionalised under neoliberalism.[45]

Klein often returns to a quote from Milton Friedman: 'Only a crisis – actual or perceived – produces real change.' For Friedman and his followers, it is imperative to *use* crises. It is precisely at such moments that 'the politically impossible becomes the politically inevitable'.[46] Individuals can win crises through speculation, but those in power can grab more power by creating big shocks and mass panic. Chapter 8 of *The Shock Doctrine* is aptly titled 'Crises Work'. They worked so well that, by the 1980s, Washington no longer relied on military dictatorships to discipline poorer countries. Various structural adjustment programmes – also known as the 'dictatorship of debt' – proved to be at least as effective.[47]

The strategy of creating crises through debt was very important in Latin America. At first, large loans were given to military juntas, but even if the countries became democracies, the loans needed to be repaid. From the early 1980s onwards, they had to be paid back at increasingly high interest rates, which forced the countries to take out further loans. According to Klein, crises were built into the Chicago School model, and Friedman's theory of crises became a self-fulfilling hypothesis: 'The more the global economy followed his prescriptions, with floating interest rates, deregulated prices and export-oriented economies, the more crisis-prone the system became.'[48] With more and more crises came more and more opportunities for Friedman, his neoliberal friends and the capitalist class to export radical ideology. During the Asian crises of 1997, many felt that the situation was made so much worse by the IMF's demands for privatisation and liberalisation that the crisis is still known today as the 'IMF crisis'.[49]

A classic example of how the ruling class has used ecological crises for its own gain is Hurricane Katrina, which hit New Orleans and elsewhere in 2005. According to human geographer Neil Smith, this was a clear example of how 'disasters don't simply flatten landscapes', they

45 Klein, *The Shock Doctrine*.
46 Quoted in ibid., p. 140.
47 Ibid., ch. 8.
48 Ibid., pp. 159–60.
49 Ramesh, 'Economic Crisis', p. 94.

'deepen and erode the ruts of social difference they encounter'.[50] The devastation of schools and schoolchildren's homes was, according to Friedman, both 'a tragedy' and 'also an opportunity to radically reform the educational system'.[51] The US TV series *Treme* (2010–13) is a stunning popular culture depiction of the class and racism at the heart of how New Orleans was first broken down and then rebuilt after Katrina. Another kind of description came from congressional representative Richard Baker: 'We finally cleaned up public housing in New Orleans.' Then he added, unabashedly, 'We couldn't do it, but God did.'[52] Poor and black people died; rich and white people rebuilt the city in their image.

An example of classquake occurred in the Indian Ocean in 2004, when an earthquake created several huge tsunamis, killing some 220,000 people and leaving over 1.6 million homeless. According to a survey by Oxfam, four times as many women than men were killed in the tsunami-affected areas of Indonesia, Sri Lanka and India.[53] For many fishing families, 'reconstruction' in Sri Lanka became synonymous with renewed destruction, as the state seized the opportunity to privatise old fishing communities and beaches and convert them into tourist resorts. Herman Kuman, chairman of Sri Lanka's National Fisheries Solidarity Movement, called the reconstruction 'a second tsunami of corporate globalization'.[54]

Who wins and who loses in a crisis depends very much on the class struggle during the crisis itself, but also on how the political economy was organised before the crisis. A functioning welfare state, social safety nets and a more equal distribution of power and money act, according to M. Ramesh, professor of public policy, as a macroeconomic stabiliser during crises.[55] This helps explain why the working class did not lose in the short term during the 1970s economic crisis in the same way they did in the crises forty years earlier or forty years later. A little further down the line, however, the new neoliberal regime of accumulation would be devastating for the organised working class.

---

50 Smith, 'There's No Such Thing'.
51 Quoted in Klein, *The Shock Doctrine*, p. 5.
52 Quoted in Smith, 'There's No Such Thing'.
53 MacDonald, 'How Women'.
54 Quoted in Klein, *The Shock Doctrine*, p. 395.
55 Ramesh, 'Economic Crisis', esp. p. 92.

Causing the Crisis

Humans have created a metabolic rift under capitalism and are destroying nature. But which class does this? The immediate answer is the working class. After all, it is hardly the children of the capitalist class who are working in the mines of Indonesia or the Congo, and no billionaires are risking their lives on oil rigs off Nigeria or paving fifty-lane highways in China. It is the working class that does all this. However, it is of course not the workers who are investing or deciding that the earth should be destroyed. This is the capitalist class, and this is the class that causes ecological crises.

Behind the metabolic rift, we find people with profit motives. This was true of the transition from hydroelectric to steam power in England in the early nineteenth century, the implementation of railways and the explosion of private cars, the development of commercial aviation, the creation of a global division of labour where goods are shipped around on boats and planes – all brought about by a capitalist class (and its allies in state and politics) to secure profits.

Overproduction, increased organic accumulation and the metabolic rift, as discussed in the previous chapter, are central structures and processes in the creation of the crises of capitalism and emerge from ordinary rounds of capital accumulation. These 'ordinary' rounds are always articulations of class society. Conventional rounds of capital accumulation are only possible through exploitation, accumulation by dispossession and destruction of nature; it is something a class can only do in relation to another class. Each individual crisis can be broken down into a million puzzle pieces, but this does not shake the general conclusion: the root cause is found within capital accumulation itself and this always comes with a class character.

One aspect that can make it difficult to see that economic crises are caused by a class is, first, that 'everyone' wants to become a capitalist. Under capitalism, there are (usually) no formal or legal rules preventing the poor from becoming rich or workers from becoming capitalists. Secondly, large parts of the working class can see an opportunity in particular booms. Before the 1930s crisis, there were an estimated 1.5 million 'small people' in the US who regularly studied stock market listings and who, according to Štajner, were 'running in a frenzy of speculative expectation of easy money to buy or to sell, to get rich or go broke'.[56] This phenomenon did not exist in the same way in the 1970s but reappeared before 2008 when real estate speculation, according to Adam Tooze, had become a 'mass sport'.[57] This still does not mean that the working class is causing the crises. For one thing, it is precisely through desperate attempts to *become capitalists* that they – to a limited extent – contribute to bubbles and booms. Secondly, the heavy and daily decisions to accumulate capital are in the hands of the capitalist class.

With the climate crisis, the same argument is easier to see, because a handful of capitalists are clearly so much more culpable than everyone else. Since 1965 – which is about when the science of climate change first became available to both politicians and the oil industry – twenty companies have contributed 35 per cent of all energy-related carbon dioxide and methane worldwide. Since 1988 – when 'everyone' knew what climate change was all about – twenty-five companies and states have accounted for half of all industrial greenhouse gas emissions. Between 1988 and 2015, 100 companies accounted for 70 per cent of global emissions.[58]

Richard Heede of the Climate Accountability Institute has carried out a historical study showing that 63 per cent of cumulative global emissions of industrial carbon dioxide and methane between 1751 and 2010 can be traced to ninety entities: fifty owned by private capitalists, thirty-one that are state-owned enterprises and nine that are nation-states.[59]

The fact that the climate catastrophe is the consequence of a series of concrete decisions by the capitalist class and top politicians does not mean, of course, that these people – as individual humans – are evil. But if owners and managers did not put profit before nature, they themselves would never have made it up the corporate hierarchy. If they do not put money before climate, the company will go bankrupt. When the

56 Štajner, *Crisis*, p. 148; see also Galbraith, *The Great Crash*, chs 2–5.

57 Tooze, *Crashed*, p. 66.

58 See, e.g., Paul Griffin, *CDP Carbon Majors Report 2017*, CDP Report, July 2017; Carl Folke et al., 'Transnational Corporations and the Challenge of Biosphere Stewardship', *Nature Ecology and Evolution* 3, no. 10 (2019), pp. 1396–403; Climate Accountability, 'Accounting for Carbon and Methane Emissions', climateaccountability.org, 2021.

59 Richard Heede, 'Tracing Anthropogenic Carbon Dioxide and Methane Emissions to Fossil Fuel and Cement Producers, 1854–2010', *Climate Change* 122 (2014), p. 229.

protection of people or nature threatens profit, companies must put profit first.

Discussions about the causes of the climate crisis focus less on ownership, production and investment; more often they are about consumption. And new reports keep coming out showing what an absurd world we live in. Between 1990 and 2015, the richest 1 per cent of the world's population was responsible for twice as much carbon dioxide emissions as the poorest half of the world's population.[60] Emissions from global aviation increased by 32 per cent between 2013 and 2018, but over 80 per cent of the world's population has never flown.[61] And Sweden? A 2020 report on individual consumption emissions between 1990 and 2015 shows that, while the bottom half of income earners reduced their emissions by 16 per cent, the richest 10 per cent increased their emissions by 1 per cent, and the richest 1 per cent increased emissions by 11 per cent.[62] Meanwhile, the lifestyles and neighbourhoods of the richest are constantly being greenwashed. Ironically, affluent neighbourhoods such as Västra hamnen in Malmö and Hammarby sjöstad in Stockholm have a reputation for being the most sustainable in the country and even the world. But this is not the case. It is in the nature of things that the poorest have the smallest carbon footprint, but no one would think of giving ecological prizes and awards to poor and stigmatised neighbourhoods like Rosengård or Rinkeby.[63]

However, the question of class and consumption is complicated by the following: although the richest consume extreme amounts more than ordinary workers, there are so many more workers. This means that, although the working class has not caused the crisis, working-class lives in the Global North will also need to change dramatically if we are to keep global warming below 1.5 degrees. But the classes consume differently.

60 Tim Gore, *Confronting Carbon Inequality*, Oxfam Report, 21 September 2020.

61 Lizzy Gurdus, 'Boeing CEO: Over 80% of the World Has Never Taken a Flight. We're Leveraging That for Growth', cnbc.com, 7 December 2017; Damian Carrington, '1% of People Cause Half of Global Aviation Emissions: Study', *Guardian*, 17 November 2020.

62 Mira Alestig and Robert Höglund, *Svensk klimatojämlikhet. Behovet av en rättvis omställning*, Oxfam media briefing, 8 December 2020.

63 See Ståle Holgersen and Anna Hult, 'Spatial Myopia: Sustainability, Urban Politics and Malmö City', *International Journal of Urban Sustainable Development* 13, no. 2 (2020), pp. 159–73.

The working class generally consumes for survival, and only occasionally for luxury. The world's richest spend almost all their income on luxuries; the proportion used for survival is for them vanishingly small. The richest consume both quantitatively (they consume more) and qualitatively (they consume unnecessary luxuries) differently from the poor.

Crises Are Caused by One Class, but Another Must Pay the Price

Our general conclusion from this chapter is that the crises are caused by the capitalist class, but mainly affect the working class. Behind this simple conclusion, we find a highly complex reality. These are two heterogenous classes and there are many interesting things happening within these classes.

This conclusion is of a different nature from what we normally encounter in research literature on class. Here, emphasis is often on inequality within the working class. Swedish class analyses, for example, focus very often on differences between workers (*arbetare*) and civil servants (*tjänstemän*), which creates some oddities, such as nurses, preschool teachers and social workers not being categorised as working class. If we use Swedish public statistics and bunch together 'workers' (37 per cent), 'lower civil servants' (12 per cent) and 'middle civil servants' (23 per cent) with the unemployed (since most of the unemployed belong to the working class), then about 75 per cent in Sweden belong to the broad working class.[64] Then we can remove workers with extremely high wages or so much (often inherited) money that they should not really need to work and therefore should rather be placed in the petty bourgeoisie or capitalist class.

Approximately 10 per cent in Sweden are 'entrepreneurs', and this must also be broken down. Therborn assigns about 8 per cent of the entrepreneurs to the petty bourgeoisie, and the bourgeoisie to about 1 or 1.5 per cent.[65] Even when we look at income, it is the 1 per cent that

---

64 Göran Ahrne, Niels Stöber and Max Thaning, 'Klasstrukturen i Sverige', in Daniel Suhonen, Göran Therborn and Jesper Weithz (eds), *Klass i Sverige. Ojämlikheten, makten och politiken i det 21:a ärhundradet* (Lund: Arkiv, 2021), p. 27. Statistics include employed workers aged sixteen to sixty-four, data from 2014–15.

65 Therborn, *Kapitalet, överheten*, p. 67. Figures from 2017. See also Martin Gustavsson and Andreas Melldahl, 'The Social History of a Capitalist Class: Wealth

stands out. According to sociologist Göran Ahrne et al., the difference between workers and white-collar workers is 'limited, to say the least', in comparison with the way income distribution looks when we include the top 1 per cent.[66]

For our purposes, we will work on the hypothesis that the capitalist class is about 1 to 2 per cent of the Swedish population. These are people who have either enough employees or enough capital to successfully accumulate, with some strategic planning. Within the capitalist class, there is usually a much smaller group with extremely high levels of social power – which, in many countries, can almost dictate the crisis policies of states. As Therborn reminds us, family sociology is often needed to analyse the capitalist class in detail – in concrete cases, in concrete countries.[67]

Questions of delimitation and classification in class analysis are always difficult. If wage-earners with wages so gigantic that they could live by accumulating capital should be included in the bourgeoisie, how high are these wages we are talking about? An entrepreneur who employs their spouse may be classified as a capitalist by some definitions, but really has no power over the development of society: How many employees does that require? If class is also about dominance over workers, how much power over others is needed? These questions present real challenges, and many seem to believe that the best answer is simply to add another class. The magical 'middle class' is particularly popular: magical because it comes in endless forms and definitions, very often without being problematised. But no matter how many classes you add to the analysis, the actual demarcations between classes remain problematic.

We will not go into an exhaustive review of the debate here – that would take up the rest of the book. Nor will we delve into Swedish figures, since our book is situated in Sweden but is not solely about Sweden. Once you have opened the discussion on categorisations, it is easy to get lost. Class is much more than statistical categorisations. Our starting point is also quite different, and we should stick to it: class is a

---

Holders in Stockholm, 1914–2006', in Olav Korsnes et al. (eds), *New Directions in Elite Studies* (New York: Routledge, 2018), p. 182.

66 Ahrne et al., 'Klasstrukturen', p. 45, our translation.

67 Therborn, *Kapitalet, överheten*, pp. 83–92.

relation and the basic dynamic is between a class that exploits and dominates another class, and nature. There are two large heterogeneous and internally differentiated classes: a capitalist class of about 1 to 2 per cent that is the core of a broader bourgeoisie, and a working class that is the vast majority. The first is strong in terms of power, the second in terms of numbers. With this we return to the crises, and we start with our rulers.

Capitalists in Crisis

Some capitalists cause the crises more than others. One approach is to divide the capitalist class into different *class fractions*, such as finance capitalists, industrial capitalists, merchant capitalists, etcetera. The class can also be divided into subgroups based on size, whether ownership is passive or active, how owners relate to states and their employees, how they are organised, personal networks and so on.

Particularly with the climate crisis, we see a distinct class fraction that stands out. Even if all of capitalism is mired in oil, gas and coal, we still need to emphasise how the fossil industry – or what Malm calls the *original accumulation of fossil capital* – is a class fraction whose historical task is to supply fossil fuels to everyone else in the market; its profits come from the production of these particular products.[68] The activity of the class fraction has been highly profitable, the commodity has had a particular geopolitical importance and the industry has had relatively few actors who have been able to cooperate in, for example, cartels, partly due to the peculiar nature of the commodity. Today, the power of this class fraction could be threatened by renewable energy sources. Or not. What is certain is that this fraction is not giving up without a fight.

Class fractions also play a role in economic crises. Under neoliberalism, the financial sector was central both in creating a more crisis-prone economy and in dictating the terms when crises were resolved. The financial sector is often considered the fraction that *must* be saved in order for the whole system to be kept alive. The question of class fractions is also central to who must be sacrificed. The typical example is

68 See Malm, *Fossil Capital*, pp. 320–6, 355–61; Malm and the Zetkin Collective, *White Skin, Black Fuel*, pp. 15–18. On class fractions, see, e.g., Nicos Poulantzas, *Political Power and Social Classes* (London: New Left Books, 1973), pp. 84–5, part 4, ch. 4.

how Thatcher, in the 1980s, sacrificed industry in the north of England for the benefit of the financial sector in London. Real power struggles within the capitalist class during a crisis can play out across multiple dimensions. At one level, it is all against all; then, it is various alliances and monopoly-like groups against others; then, it is fraction against fraction, and even city against city, region against region or nation against nation.

One might think that the oil and gas sector – and the class fraction rooted in the *original accumulation of fossil capital* – would not need state support. But, in 2018, states subsidised fossil fuels to the tune of over $427 billion.[69] A prudent first step to saving our planet would be to stop paying corporations to take it apart. Measured by revenue, the industry known as 'global oil and gas exploration and production' is the world's fifth largest. Others on that list are life and health insurance, pension funds and the car industry. The world's largest sectors are killing the planet with one hand and making money from our misfortune and illness with the other.[70]

The capitalist class also consumes most. The general trend is that the higher the wages and the greater the wealth, the bigger the carbon footprint. Wealth cannot simply be reduced to ownership, even if they are related. We therefore also need to build an analysis with categories such as income, education, type or price of housing and health, and then make a hierarchical order. An analysis that *only* does this is often called stratification analysis.[71]

It is worth pausing for a moment on climate and consumption, as this has become a minefield with political implications. The American geographer Matt Huber has formulated a harsh and apt critique of the consumption perspective when he asks, 'Whose carbon footprint

---

69 IEA, *Energy Subsidies: Tracking the Impact of Fossil-Fuel Subsidies*, IEA Report, 2019; Angus, *Facing the Anthropocene*, p. 170.

70 IBISWorld, 'Global Biggest Industries by Revenue in 2021', ibisworld.com, 2021.

71 For argument that all class analysis is stratification analysis, cf. Peter Saunders, *Social Class and Stratification* (London: Routledge, 1990). For critique, see James Stolzman and Herbert Gamberg, 'Marxist Class Analysis versus Stratification Analysis as General Approaches to Social Inequality', *Berkeley Journal of Sociology* 18 (1973), pp. 105–25. On Marxist and non-Marxist class theories, see Erik Olin Wright, 'Understanding Class: Towards an Integrated Analytical Approach', *New Left Review* 60 (2009), pp. 101–6; Wright, *Understanding Class*; Rosemary Crompton, *Class and Stratification* (Cambridge: Polity, 2008).

matters?'[72] On websites such as climatehero.me/en, you can calculate your carbon footprint in five minutes by entering whether you drive or eat meat and how you live. A CEO of a major airline company can, during working hours, decide to expand the company's fleet with hundreds of new planes and thousands of new flights, thereby increasing both the company's turnover and his own bonus. Then he cycles home to eat a vegan burger for dinner in his eco-house. Calculations on carbon footprints focus on the bike and the burger, never what an individual does *as a capitalist*. Huber is right: what they do as a capitalist is immensely more important than what they do as a consumer.

Huber also argues that the focus on consumption gives the climate movement a definite class character because this has historically appealed to a particular type of at least partially privileged people. Not just anyone can afford to buy more ecologically and thus 'be' ecological. It is truly absurd to tell people who cannot afford to buy Christmas presents for their children that they must consume less. Huber's main argument is crucial: the climate debate must be more about production.

Then Huber walks a few steps too far. Even though ownership is more important than consumption, that does not mean consumption is not also important.[73] This will complicate the analysis, but we cannot ignore consumption for several reasons: mass consumption will always play a central role in capitalism; grotesque international inequalities among workers in different countries force us to think globally about said questions; class struggles have never existed *only* at workplaces, but, in societies organised on the principle of mass consumption – where the rich

72 Matt Huber, *Climate Change as Class War: Building Socialism on a Warming Planet* (London: Verso, 2022).

73 Huber argues in *Climate Change as Class War* that he has a Marxist understanding of class (pp. 3, 20), where classes are defined as standing in antagonistic relations to each other. But some of his core arguments are articulated within a Weberian framework (e.g., when distinctions between the professional-managerial class [PMC] and the working class are based on advantages within labour markets, see p. 6), a Bourdieusian framework (e.g., emphasising the role of 'credentials', see p. 5, and even culture and taste as distinctions between 'workers' and PMC) and psychoanalysis (as the focus on consumption is apparently driven by 'carbon guilt' among PMCs who want to blame themselves for climate change, see p. 29). On this last point, I think it is true that many people whom Huber defines as PMC want to blame themselves. But the opposite is also true: many 'PMCs' want to ignore consumption in order to legitimate personal meat consumption and flying.

consume luxuries while others cannot pay the rent – class and class hatred will always be expressed also through questions of consumption. Since luxury consumption normally comes from exploiting workers and destroying nature, the issue of consumption contains political potential: emphasising luxury consumption can help unite the environmental and socialist movements. Environmental activists trying to stop private jets are doing exactly this. The ruling class are destroying the planet as both capitalists *and* the richest among us. There is no reason to let them get away with any of it. We need class struggle both against the owners of means of production, and against owners of luxury yachts.

Workers in Crisis

As with the capitalist class, there are also major internal differences within the broad working class. About 3.5 billion of the world's 7.8 billion people are part of what the ILO defines as the 'global workforce'. In addition, there are 2.2 billion *outside* the labour force (almost all of whom belong to working families and thus the working class), tens of millions in prison (sometimes used in slave labour), about 150 million child labourers, and perhaps a billion farmers (the vast majority of whom are small farmers and could therefore also be included in the broad global working class). The global labour force also includes some 470 million unemployed or involuntarily part-time workers.[74] Of all people in the global workforce, 53 per cent are wage workers. This figure is considerably lower than in the Global North – for example, in Sweden it is 90 per cent – which can be explained by large informal sectors and gigantic numbers of precarious 'self-employed' in many poor countries.[75]

Of course, the billions of workers in the world are affected very differently by the crises of capitalism. From a stratification analysis based on income, we can say, as a general tendency, that both economic and ecological crises hit harder the poorer someone is. Lower income means less of a buffer to face the crisis – but, even here, there are exceptions as

74 See ILO, *World Employment and Social Outlook Trends 2020*, International Labour Organization, Flagship Report, 2020, pp. 2, 5, 20, 24; Yates, *Can the Working Class*, ch. 1. Yates uses 2018 figures.

75 World Bank, *Wage and Salaried Workers, Total (% of Total Employment) (Modelled ILO Estimate)*, 2021. ILOSTAT database; ILO, *World Employment*, p. 2; Yates, *Can the Working Class*, pp. 18, 36–45.

levels can vary between sectors, fractions, geographical areas and so on. We also need to consider job type and housing situation to understand how crises impact differently. Such a focus is often associated with Max Weber and his definition of class as groups sharing the same life situation in the labour, housing and commodity markets. In this tradition, the working class is those *excluded* from for example ownership of businesses, from education (the labour market) or from the ability to buy their home (the housing market).[76]

Workers certainly lose in economic crises as individuals, but do they also lose as an organised class? One perspective that can give us a hint is how unions are affected. The general conclusion from research is that unions usually end up in a more difficult situation during crises and (threats of) increased unemployment and lower growth, but the picture is complex.[77] The Great Depression in the United States reappears here as an interesting case. It is often argued that social misery made people more sympathetic to the struggles of poor workers, which, in turn, benefitted the labour movement. Historian Steve Fraser argues that the case was often the opposite. The workers' cries for help during deep social distress were not answered at all. When the crisis hit in 1929, many unions immediately lost members, forcing them to cut back on organising and in-house training. This changed radically from 1933 to 1934, but, by then, the worst crisis as *shock* and sheer desperation was over. 'It is reasonable to think', writes Fraser, 'that the economy's improvement gave people the courage, the optimism and the material leverage with which to fight back.'[78] In 1934, 1.5 million people went on strike, and with rent strikes and massive protests by the unemployed and

76 For further discussion, see Wright, *Understanding Class*, p. 6; Crompton, *Class and Stratification*, ch. 3; Saunders, *Social Class*; Richard Breen, 'Foundations of a Neo-Weberian Class Analysis', in E. O. Wright (ed.), *Approaches to Class Analysis* (Cambridge: Cambridge University Press, 2005), pp. 31–50; see also Weber, *Economy and Society*, pp. 302–7, 926–40; Göran Therborn, *What Does the Ruling Class Do When It Rules?* (London: Verso, 2008), pp. 140–3.

77 See Dan Cunniah, 'Preface', *International Journal of Labour Research* 1, no. 2 (2010), pp. 5–7; Frank Hoffer, 'The Great Recession: A Turning Point for Labour?', *International Journal of Labour Research* 2, no. 1 (2010), pp. 99–117. For trends in Norway showing that union membership historically has declined (or grown more slowly) during crises, see Harald Berntsen, 'Kapitalismens kriser og arbeiderbevegelsens svar', *Sosialistisk Framtid*, 1 March 2021.

78 Steve Fraser, 'American Labour and the Great Depression', *International Journal of Labour Research* 2, no. 1 (2010), p. 15, see also pp. 9–24.

agricultural workers, the labour movement was on the offensive. This happened when the worst of the shock had subsided, but when the larger hegemonic crisis was still unresolved. And we must stress that this radical wave did not come out of nowhere; it was planned and executed by organised socialists.

In the 1970s, trade unions also lost out in the crisis, especially in the longer term. Belgian economist Bob Hancké argues that French unions lost many *more* members than German ones, and that this was because unions that relied on a combination of centralised and decentralised organisation fared better than those labour movements that lacked strong local union capacity.[79] If any strong trade union movement could have used the crisis as an 'opportunity', it should perhaps have been the Swedish one in the face of the 1990–94 crisis. But it did not. Ingemar Lindberg and Magnus Ryner have shown that, more than anything, the crisis provided an opportunity for capital to continue on its track and accelerate the erosion of the Swedish model.[80]

Trade unions were marginalised across the eurozone during and after the 2008 crisis.[81] This does not mean that unions or socialist movements did not play any role. Unions can alleviate problems: a study of twenty-one countries in Central and Eastern Europe found that union members were less likely than non-members to lose their jobs during the 2008 crisis, even though the price was often a pay cut.[82] We also do not know how the crises would have unfolded without trade unions and socialist movements.

A common focus in economic crises is the instability that arises when the so-called middle class gets into trouble and starts to 'sink'.

79 Bob Hancké, 'Trade Union Membership in Europe, 1960–1990: Rediscovering Local Unions', *British Journal of Industrial Relations* 31, no. 4 (1993), pp. 593–4; for a review of research up to 1993 see pp. 594–7.

80 Ingemar Lindberg and Magnus Ryner, 'Financial Crises and Organized Labour: Sweden 1990–94', *International Journal of Labour Research* 2, no. 1 (2010), p. 27. For similar discussions on Japan and South Korea and their 1990s crisis, see Hoffer, 'The Great Recession', p. 102.

81 See Philip Rathgeb and Arianna Tassinari, 'How the Eurozone Disempowers Trade Unions: The Political Economy of Competitive Internal Devaluation', *Socio-Economic Review* 20, no. 1 (2020), pp. 323–50.

82 Artjoms Ivlevs and Michail Veliziotis, 'What Do Unions Do in Times of Economic Crisis? Evidence from Central and Eastern Europe', *European Journal of Industrial Relations* 23, no. 1 (2017), pp. 81–96.

We will not use 'middle class' here as the term mostly creates ambiguity and vagueness, but the phenomenon the phrase 'the falling middle class' is trying to describe is relevant. In times of crisis, even workers who used to have relatively high wages and who thought they had secure jobs can suddenly find themselves in much more precarious situations. There can also be entrepreneurs who go bankrupt and may be forced into wage labour. Small business owners might find themselves in a foxhole: as capitalists they may have to increase the exploitation of their handful of employees because they are in debt to banks demanding immediate reimbursement.[83] While the interwar period is often held up as the exemplar of the reactionary consequences of such developments, a generation of educated young people in southern Europe after the 2008 crisis showed that they can be a progressive force, even when the prospects for the future seem very difficult.

When analysing the complex class character of crises, we need to use a mix of approaches. The case of Hurricane Katrina is an interesting example of how we need to see rich capitalists as *both* rich *and* capitalists. It was as *rich people* (in a racist society) that they were able to escape the immediate disaster and get out of New Orleans, get medical care and alternative housing; but it was as *capitalists* that some of them were able to use the crisis to rebuild the city (now even more clearly in the self-image of the capitalist class).

From the hypothesis that crises are caused by one class and that the bill is paid by another, we can return to the conclusion of the previous chapter and say that crises are necessary for the capitalist system but a danger to the world's workers. What benefits the system does not necessarily benefit a majority of the people who live in it. If crises are creative destruction, it is primarily creative for the system and primarily destructive for the broad working class. If crises are a necessary evil, they are necessary for the system and evil for us.

In its Greek origin, the word 'crisis' was associated with judgement and decision. In the Greek *polis*, crisis referred to the process of bringing an action to litigation or the passing of a judgement, and in Aristotle,

83 On contradictory class positions, see Erik Olin Wright, *Classes* (London: Verso, 1997), ch. 2; Crompton, *Class and Stratification*, ch. 6.

crisis is also associated with a 'just judgment'.[84] It would be great if the crises of capitalism were really a court of justice, where we all could stop for a moment and reflect together on how the political economy should be organised. But our capitalist crises are not the fair judges, courtrooms or trials of history. Rather, they are the arena of power. They are sites of unjust reproduction. That class struggle intensifies during crises of capitalism may sound like a dream to the left, who might be more than happy to welcome some extra class struggle. But most of this is nothing to cheer about. This is class struggle from above, subtly and quietly, often with murderous efficiency. Rather than being memento mori, the crises have kept capitalism alive; but for workers and the poor, the crises have become concrete reminders that life does indeed have an end.

No One Is Punished, but Murder It Remains

A bank robber caught by the police goes to jail; if you punch someone in the face you get your punishment; and parking incorrectly results in a fine. Conversely, when millions of people lose their jobs or homes because of man-made crises, or when hundreds of millions have to flee because of anthropogenic climate change, no one is held to account. The causes and consequences of crises may not have as direct a link as when one person strikes another, but crises are man-made *and* they have terrible human consequences. That no one is punished is a mystery.

That absolutely no one is ever punished after a crisis is not entirely true. After 2007–08, a total of twenty-seven bankers were jailed in the wake of the crisis; most served only a few years or months, half of them in Iceland and only one in the US. None were Wall Street executives. According to the Better Markets organisation, there was evidence of illegality, but the Wall Street executives got off scot-free because of the almost unprecedented level of power and influence of the financial sector.[85]

84 Brian Milstein, 'Thinking Politically about Crisis: A Pragmatist Perspective', *European Journal of Political Theory* 14, no. 2 (2015), pp. 154–5; Janet Roitman, 'Crisis', in *Political Concepts: A Critical Lexicon* 1 (2012).

85 Better Markets, 'The Cost of the Crisis $20 Trillion and Counting', bettermarkets.org, 2015, p. 5; Laura Noonan et al., 'Who Went to Jail for Their Role in the Financial Crisis?', *Financial Times*, 20 September 2018; Cervenka, *Vad gör*, pp. 132–5. Credit

Critics have long discussed how capitalism kills. Development economist and former Chicago economics dissident Andre Gunder Frank wrote a letter to Milton Friedman over forty years ago accusing him of economic genocide. Moscow academic Vladimir Gusev concluded after Russia's economic reforms in the 1990s that the 'years of criminal capitalism have killed off 10 percent of our population'.[86] In radical circles, we may hear that neoliberalism kills or capitalism kills, but often it feels more like phraseology.

Considering that the capitalist class both benefits from and causes the crises, and given they are the rulers of the world, it's intriguing how anonymous this class typically remains. One reason for this is the nature of the killing – it is never the rich themselves who kill anyone – to which we will return below. When liberals discuss inequality and power, they assume that private ownership of capital is a human right that should not be questioned, and this assumption erases the category of the capitalist class. The class is invisible in other ways as well. Public discourse tends to focus on the newly rich, and the media love entrepreneurs who have hit the jackpot and gone straight onto the list of the world's very richest. This tendency masks all the things that change at a much slower pace: all the ownership that is passed down and the daily reproduction of the ruling class.

Another reason for the magical invisibility cloak around the capitalist class is that focus within class theory from the 1980s onwards shifted from a macro-foundation (focusing on division of labour, large dominant classes) to a micro-foundation (emphasising individuals' acquisition of 'assets', 'capital' or 'resources').[87] The general obsession with cultural differences between wage-earners with a focus on taste and aesthetics, and/or an extreme obsession (fetishisation) with the so-called middle class further obscures the fact that some actually own capital. The fact that it can be so difficult to see who has created the crises can partly also be explained by capitalism's inherently undemocratic nature:

---

rating agency Moody's agreed in January 2017 to pay nearly $864 million to the US government as a penalty for giving inflated ratings to risky housing packages before the 2007–08 crisis: see Associated Press, 'Moody's to Pay Nearly $864 Million to Settle Claims It Inflated Ratings', CBS News, 14 January 2017.

86 Quoted in Klein, *The Shock Doctrine*, p. 238.

87 See, e.g., Crompton, *Class and Stratification*, pp. 109–16; Bensaïd, *Marx for Our Times*, ch. 5; Gustavsson and Melldahl, 'The Social History', p. 78.

when no one has elected the capitalists to run the economy, it is easy to assume that no one can be held accountable.

The capitalist class has disappeared from view in the scientific world, but their disappearance is even more evident in the media and politics. While the general abstraction 'capital' appears everywhere, capital's owners are rarely discussed. Capital seems to be something that moves with its own momentum and will, but in reality it is by definition impossible to talk about capital without talking about class. Where the working class or the so-called middle class is presented as flesh-and-blood individuals with different fates, resources and histories, the capitalist class is usually portrayed in general categories such as 'capital', 'finance' or 'investment'. This is evident in studies of the city and gentrification: people moving in and out of certain areas are referred to as middle and working class and described as people with agency (at least the so-called middle class). In contrast, the capitalist class is usually completely hidden, and, when it appears, it is through categories including financiers, property owners, builders, contractors, speculators and developers, but almost never as a class.[88]

The working class consists of people and actors, whereas the capitalist class consists of abstractions and structures. In socialist literature, it is not uncommon to read that the working class (human) is opposed to capital (abstraction). By comparison, we seldom read that the capitalist class stands against labour.[89] In education, economics is taught in business school (objective capital), while class has been reduced to some sub-discipline in sociology (subjective people). In the news, we first see descriptions of the human misery among the working class, then comes the *economic news* with reports on how the stock market and corporations are doing. In the political sphere, there are *ministers of business* and *finance* to make general decisions and make the overall market work, and *ministers of social affairs* to make sure that real people further down the social hierarchy do not die of hunger. In mainstream media, when capitalists are portrayed as individuals, they appear as celebrities, rarely as people with social power, never as a class. The culpability of the

88 See Ståle Holgersen, 'Lift the Class – Not the Place! On Class, Urban Policies in Oslo', *Geografiska Annaler: Series B, Human Geography* 102, no. 2 (2020), pp. 135–54.

89 See also Immanuel Wallerstein, 'The Bourgeois(ie) as Concept and Reality', in Étienne Balibar and Immanuel Wallerstein, *Race, Nation, Class: Ambiguous Identities* (London: Verso, 1991), p. 143.

capitalist class is hidden behind hegemonic liberal discourses. But there is also something in the very nature of the crises that complicates the issue.

Can Structures Cause Misery and Death?

In both ecological and economic crises, there is cheating and criminality, but the crises themselves would not disappear even if everyone obeyed the law. Here we cannot avoid the difficult exercise of trying to see crises as both structures and actors at the same time. From a structural perspective, we can see that crises unfold beyond the day-to-day control of individuals, firms or states. In a preface to the first edition of *Capital*, Marx also writes: 'My standpoint . . . can less than any other make the individual responsible for relations whose creature he remains, socially speaking, however much he may subjectively raise himself above them.' In the *Grundrisse* he writes that 'individuals are now ruled by *abstractions*'. 'The Roman slave', Marx writes in *Capital*, 'was held by chains; the wage-labourer is bound to his owner by invisible threads.'[90] There are indeed iron laws within capitalism that operate beyond the control of individuals, the capitalist class and states. Crises are produced independently of the intentions of individuals, and the ruling class absolutely rules the world *with* abstractions, and we are ruled *by* abstractions. There are impersonal and abstract relations and dynamics to which everyone must relate, and which reproduce capitalism on a daily basis.[91] According to the Danish philosopher Søren Mau, we must therefore avoid reducing the *power of capital* to the *power of capitalists.*[92] This is correct. But the argument should not be taken too far in the other direction either. The power of the actual capitalist class must never be reduced to mere impersonal abstractions.

Markets or laws of value are not magical or supernatural forces that can be reproduced without humans, even if they might seem so. No abstractions have ever killed anyone – at least not on their own. In our

90 Karl Marx, *Capital*, vol. 1 (London: Penguin, 1976), pp. 92, 719; Karl Marx, *Grundrisse* (New York: Vintage, 1973), p. 164.

91 Søren Mau, '"The Mute Compulsion of Economic Relations": Towards a Marxist Theory of the Abstract and Impersonal Power of Capital', *Historical Materialism* 29, no. 3 (2021), p. 8.

92 Ibid., p. 13.

investigation of the deepest level of abstraction of crises, we found not only exchange-value, but also use-value, that constant reminder that we are real people on a very real planet. Socialists cannot run around chasing abstractions. No revolution has ever been made against pure abstractions. We cannot escape the fact that we have to deal with both abstractions and actual people at the same time, and no advanced theory can conjure this tension away.

If we are *also* governed by abstractions, can we put an abstraction on the defendants' bench? How do we approach the question of guilt when the production of human misery is embedded in structures? In Engels's description of poverty in English cities in the 1840s, we get an entry point that is relevant even today:

> When society places hundreds of proletarians in such a position that they inevitably meet a too early and an unnatural death, one which is quite as much a death by violence as that by the sword or bullet; when it deprives thousands of the necessaries of life, places them under conditions in which they *cannot* live – forces them, through the strong arm of the law, to remain in such conditions until that death ensues which is the inevitable consequence – knows that these thousands of victims must perish, and yet permits these conditions to remain, its deed is murder just as surely as the deed of the single individual; disguised malicious murder, murder against which none can defend himself, which does not seem what it is, because no man sees the murderer, because the death of the victim seems a natural one, since the offence is more one of omission than of commission. But murder it remains.[93]

Even if it is people who kill – consciously or unconsciously, intentionally or unintentionally – the issue remains complicated. If we are to take the argument further, we must again distinguish between the ecological and economic crises. In economic crises, individual criminal actors can be punished – speculators who contribute directly to crises, banks that cheat their customers out of loans, politicians who create (pseudo)crises to promote an ideology, or rating agencies that give top ratings to investments they know are worth little. Punishing these

93 Engels, *The Condition of the Working Class*, pp. 95–6.

crooks might be good and fair in itself, but it would not prevent future economic crises.

With climate crises, the picture is a little different. For example, when the Decolonial Atlas publishes a map of the top hundred people killing the planet, this is both important and insufficient.[94] It is important because it reminds us that the earth is not an elderly person now lying on her sickbed. As the American folksinger Utah Phillips so famously put it: the earth is not dying, it is being killed, and the people who are killing it have names and addresses.

More and more voices are now arguing that ecocide must be included in the International Criminal Court (ICC).[95] The ICC was established in 2002 in The Hague precisely to punish perpetrators of genocide, crimes against humanity and war crimes. Why not climate change as well? Or is mass murder of the world's poorest legitimate as long as it is done via the atmosphere and not directly with weapons?

It is morally right to demand that ecocide become an integral part of the ICC. This would put the onus on companies, organisations and states that deliberately profit from (indirectly) killing others, and could act as a threat to politicians and big business – if you escalate climate change, expect to be taken to court. At the same time, there are limits. Ecocide trials can only punish the worst. Does the need to punish *someone* make us miss the fact that crises are *systemic* problems? The crises of capitalism do not happen primarily because someone does something criminal, they happen because what should be criminal is perfectly legal.

It is important to point out that 100 companies account for 70 per cent of global emissions, but this is insufficient because it hides the fact that companies often supply fuel – or other goods, or services – to the entire capitalist economy. Even if the state confiscated all the money and privileges of the hundred people doing the most to damage the planet, the problems would still persist, because others would take over their positions. This is where keeping the focus on ownership becomes key. The problems remain precisely because these 100 are not just rich individuals, but also capitalists.

---

94 The Decolonial Atlas, 'Names and Locations of the Top 100 People Killing the Planet', decolonialatlas.wordpress.com, 27 April 2019.

95 See, e.g., Dawn Allen, 'Ecocide May Actually Become a Crime Soon', legalreader.com, 7 December 2020.

When the short-term goal is to limit global warming to below 1.5 degrees, companies, organisations or states that have deliberately lied, falsified information or misled populations to legitimise continued greenhouse gas emissions or continued deforestation should be brought to justice. This could be an important tool in the struggle. Such an emphasis risks, at the same time, winding up in endless discussions about laws, rules, bureaucracy, procedures and justice.[96] The deeper metabolic rift can hardly be confronted in a courtroom within a capitalist state apparatus.

The most important action required is the same in both economic and ecological crises: to put the capitalist class on trial. Ending the era of crises can only be achieved through a class struggle that manages to navigate between structures and agents.

Class Struggle in Times of Crisis

Generally speaking, we can point to three different ways in which the capitalist class daily and often quietly rules over the working class.

First, exploitation. Here, we understand exploitation rather narrowly, as what happens when wage labour produces surplus value. Uncovering how this works was one of Marx's major contributions to world history, and the theory still stands: workers produce more value than they are paid, and this surplus value must partly be reinvested and thrown into new rounds of capital accumulation. Exploitation, as defined here in its Marxist sense, might seem like a small component of a big world. But it has disproportionate consequences and is core to understanding how capital can accumulate and how the economy can grow.

Second, accumulation by dispossession. People can be oppressed and abused in all sorts of ways that are not exploitation as described above. Small farmers, poor workers in the Global South, slaves, indigenous peoples and many others can be dominated and tormented by the ruling class even outside wage labour. Through various forms of privatisation, rentierism, debt creation, land ownership, real estate speculation and more, capital – which is based on inequality of power and ownership

96 For critical discussions, see Laura Pulido, Ellen Kohl and Nicole-Marie Cotton, 'State Regulation and Environmental Justice: The Need for Strategy Reassessment', *Capitalism Nature Socialism* 27, no. 2 (2016), pp. 12–31.

– can be accumulated. Thievery does not create more value, but it has always played a central role in capitalism. Fraud and gambling, swindling and cheating have all contributed to both capitalism's advances and its crises. On a global scale, this has occurred through the plundering of colonies and through imperialism.

It is now the dominant view in critical theory to see original accumulation as something that did not end when capitalism was established. On the contrary, it continues in new forms as new people, sectors and geographical locations are drawn into capitalism. The pioneer of this theory-building was Rosa Luxemburg, with her analyses of how capitalism needed to continually open new markets in order to expand. In the early 1970s, Mandel wrote that original capital accumulation and capital accumulation through surplus value production are not only successive phases in economic history, but also two simultaneous economic processes. But while the historical original accumulation describes a capitalism *in the making*, current processes take place within already established capitalist societies.[97] It was David Harvey who really captured and popularised the phenomenon with the term 'accumulation by dispossession', which is still the best way to grasp the phenomenon.[98]

Third, accumulation through degradation of nature. Just as the original accumulation did not stop tormenting people once capitalism was established, the metabolic rift is still also thriving. The destruction of nature under capitalism is, of course, implicit in both exploitation and accumulation through dispossession, but we will nevertheless highlight this as a category in its own right because it comes with distinct characteristics, and in order to emphasise that the destruction of nature is a class relation.

The destruction of nature becomes a class relationship as one class dominates – and owns – nature, while another pays the consequences of this dominance. Dominating nature thus becomes a way of reproducing

97 Mandel, *Late Capitalism*, pp. 46–7.

98 For Luxemburg and Harvey this is related to underconsumption and overproduction as a theory of crisis: Rosa Luxemburg, *The Complete Works of Rosa Luxemburg*, vol. 2, *Economic Writings 2*, ed. Peter Hudis and Paul Le Blanc (London: Verso, 2015), pp. 261–4, 266–7, 302–3, 329–30; David Harvey, *The New Imperialism* (Oxford: Oxford University Press, 2003), ch. 4. Nancy Fraser and Michael D. Yates call accumulation by means other than exploitation of wage labour *expropriation*: Nancy Fraser and Rahel Jaeggi, *Capitalism: A Conversation in Critical Theory* (Cambridge: Polity, 2018), pp. 39–47, 101–8; Yates, *Can the Working Class*, pp. 45–56.

the power of the capitalist class over workers. We can show this by discussing ownership, metabolic rifts, profit imperatives, exchange-value and so on, more theoretically. Or we can start pointing to historical examples, as when Malm shows how the shift from hydroelectricity to coal in the early nineteenth century was very much about controlling and gaining easy access to workers; or when environmental science pioneer Barry Commoner demonstrated in 1976 that handbags were increasingly made of plastic instead of leather despite the fact that this was far more energy-intensive, because it could lower labour costs.[99] According to O'Connor, we also see the class nature of environmental regulation when we ask who tends to oppose campaigns for a better environment. It *may* be local communities or workers who, for example, want to preserve local but environmentally destructive workplaces, but the general answer, according to O'Connor, is still the capitalist class. The demand to protect nature from exploitation, to protect workers from environmentally hazardous work, to demand democratic decisions over nature – all of these are seen by the capitalist class as interference with their very property rights.[100]

Unfortunately, parts of the left have also struggled to update their class analysis in relation to the environment. The idea that the climate crisis is a challenge we can only solve if *we all* – people, communities, companies – work together is embraced within certain left-leaning circles as well. Among some socialists, any criticism of the oil industry is seen as directly disrespectful to workers in these sectors.[101] Others argue that 'ecology is a new opium for the masses' that prevents us from focusing on the real class struggle.[102] The ecological crises of capitalism mean that we need to update class analysis, but without throwing the baby out with the bathwater. Not only are socialists who fail to integrate class and ecology irrelevant in the twenty-first century, they are missing an obvious target.

Since the capitalist class dominates people and nature in different

99 Malm, *Fossil Capital*; Barry Commoner, 'Oil, Energy, and Capitalism', talk given at the Community Church of Boston, 22 February 1976. Published on climateand captalism.com, 30 July 2013.

100 See O'Connor, 'Capitalism, Nature', p. 37.

101 Cf. Eivind Trædal, *Det svarte skiftet* (Oslo: Cappelen Damm, 2018), p. 254.

102 Alain Badiou, quoted in Kohei Saito, 'Marx's Theory of Metabolism in the Age of Global Ecological Crisis, Deutscher Prize Memorial Lecture', *Historical Materialism* 28, no. 2 (2020), p. 10.

ways – exploitation, accumulation through dispossession and accumulation through destruction of nature – resistance also takes different forms. Reading only the first volume of Marx's *Capital*, it is easy to conclude that class struggle only takes place in workplaces. But if we look at all socialist struggles since the publication of *Capital*, it is clear that class struggle takes place in many other places too. Class struggle has, for example, been very much about housing and urban issues, food issues (bread demands), social reproduction, ecology, consumption, culture and so on.

When people who are deeply affected by the consequences of capitalist crises direct their anger against the capitalist class, this is class struggle. I want to emphasise that class struggle does often occur in its 'classic' form as organised workers against some capitalists, and that labour unions and other workplace struggles will always be absolutely central for socialist movements. But class struggle in general cannot be understood – let alone succeed – unless it is also embedded in other issues. One day the French put on yellow vests and stop traffic, the next day there is a fight against fossil gas outside Gothenburg, and soon after there is Black Lives Matter in the US. These are three different struggles in three different places, all with the *potential* to be socially transformative, but whether they are actually articulated as class struggles depends on the political development of the movements.

Class and class struggle develop alongside other social power relations, and lessons from intersectional, anti-racist and social reproduction theory can help us understand how these intersect in different ways.[103] Generally, such issues take a back seat in this book, but, in reality, they interact all the time – from how capitalism is organised and thus how crises are created, to who is affected by the crises and what resistance looks like. Here, we will just stress that class struggle in our time must avoid both class reductionism (understood as everything being about class alone) and individual-centred identity politics (everything being about individual experience).

---

103 See, e.g., Tithi Bhattacharya (ed.), *Social Reproduction Theory: Remapping Class, Recentering Oppression* (London: Pluto Press, 2017); Paulina de los Reyes, Irene Molina and Diana Mulinari (eds), *Maktens (o) lika förklädnader: kön, klass & etnicitet i det postkoloniala Sverige: en festskrift till Wuokko Knocke* (Stockholm: Atlas, 2005).

Climate Crisis as Class Struggle

Fruitful discussions around class struggle and the climate crisis are often hindered due to a stark polarisation between socialist eco-modernism and degrowth currents. Matt Huber, representing the former, in *Climate Change as Class War* defines classes as standing in antagonistic relations to each other *and* he argues the working class is a separate class from the professional-managerial class (PMC), which includes the entire environmental movement.[104] There can surely be *tensions* between organised workers and people within the environmental movement, but if there are *antagonistic* relations, this means class struggle: Is there really a class struggle between the working class and the 'class' that has occupied the environmental movement? Such a belief in the antagonism between a 'progressive' working class and a 'reactionary' PMC is often mirrored by degrowth critiques of ecological imperialism. This view can be represented by degrowther Tadzio Müller who has argued that not only will industrial workers in the Global North be our enemies, 'they will be our most effective enemies'.[105] Here, discussions on class start – and often end – by pointing out that workers in the Global North have an 'imperial' mode of living.

This polarisation effaces complexity and obscures the best path forward. To understand existing and potential relations between actually existing workers and climate change, we must grasp the working class as heterogeneous and understand differences among unions, relations to class struggle outside workplaces, geographies, age, gender and much more. It is intellectually dishonest to ignore tensions between workers and climate, racism and imperialism. But it is also politically hopeless to think these tensions are so great that 'workers' – however defined – cannot or should not be subjects for stopping global warming. It remains an absolute prerequisite for eco-socialists that organised labour (often alienated by degrowth movements) and environmental movements (often alienated by left-productivists) are not only radicalised and strengthened, but also brought together.

This should not be formulated as the need to reconcile the environmental movement and 'class'. The climate movement is made up of

104 Huber, *Climate Change*, pp. 3, 20.

105 In '#123 Blow Up Pipelines? Tadzio Müller and Andreas Malm on What Next for the Climate Movement', 5 May 2021, YouTube.

people who do not own any means of production (i.e., the broad working class) and their articulated main enemy is the fossil fuel industry (i.e., a fraction of the capitalist class). *This is already class struggle.* That class consciousness is low – sometimes extremely low – among parts of the movement is indeed a problem. That problem is compounded by socialist eco-modernists and degrowthers discursively reproducing and cheering on the conflict, often with a focus on aesthetics and taste, culture and education, and often (unconsciously or not) seeing 'workers' as only (white) male industry workers. When understanding class struggle in terms of exploitation, accumulation by dispossession *and* destruction of nature, we see that climate struggle *is* class struggle.

The main socialist challenge is to reconcile the class struggle in the environmental movement with the class struggle in the workplace. This is only possible through an organised and conscious socialist movement. That bringing together the broad working class is a difficult task should not surprise us: this has been the case for two centuries.

Class struggle from below is about confronting the capitalist class, either directly (strike!) or indirectly (stop privatisations and environmental destruction). Indirect struggles have always been central to the class struggle, and here we include the crises of capitalism. Since the crises themselves seldom bring any benefits to the working class, we must *turn* the crises into class struggles from below. As the capitalist class maintains its power because of, not in spite of, the crises, we need class struggle against the crises.

Göran Therborn paraphrases Erik Olin Wright when he asks: If class is the answer, what is the question? Therborn argues that class is the answer to two questions. First, why is there abundance and poverty, privilege and misery at the same time? And second, what social forces can disrupt these conditions?[106] These are two questions that lie at the heart of critical class analysis and need all the attention they can get. In this chapter we have seen that class is also the answer to a third question: What causes the crises of capitalism?

106 Göran Therborn, 'An Agenda for Class Analysis', *Catalyst* 3, no. 3 (2019), p. 89; Erik Olin Wright, 'Conclusion: If "Class" Is the Answer, What Is the Question?', in Erik Olin Wright (ed.), *Approaches to Class Analysis* (Cambridge: Cambridge University Press, 2005), p. 180.

And we can go one step further. At one level, each capitalist stands alone – as individuals, they are indifferent to each other and benefit from others going bankrupt. At the same time, they are interdependent. 'Private interest', Marx writes in *Grundrisse*, 'is itself already a socially determined interest.'[107] Capital is a common product that can only be set in motion by the common activity of people. 'Capital is therefore not only personal; it is a social power', Marx and Engels write in *The Communist Manifesto*.[108] Class as social power is the power to shape the world – to create a 'world after its own image'.[109] The capitalist crises presuppose a certain size of a social system organised on the basis of exchange-value and having a general infrastructure. This can only be done by capitalists in a social context. Or, in other words: as a class within capitalism.

As wealthy *individuals*, capitalists can buy islands and fly private planes. They can change landscapes by constructing villas, buildings, even cities in their image. As wealthy *capitalists*, they change the world far more dramatically. It is only as a class that they produce global trade networks with infrastructures of ever-new air routes and ports; mass production that creates incredible amounts of waste; an energy system whose main purpose is to serve an exponentially growing economy; extreme poverty in some continents and extreme wealth in some neighbourhoods; and interconnected financial centres that can move money in the blink of an eye. As wealthy individuals, they can create places in their image, but it is only as the ruling class that they create a *world* in their image.

The social power of the capitalist class stands on a base of *exchange-value* and is expressed through *money* – this universal and abstract means that can build financial headquarters in Stockholm one day and invest in oil sands in Canada the next. We have seen how the crises of capitalism are due to the contradiction between exchange-value and use-value, where production is organised according to the principles of exchange-value. Now we can see how the crises would be impossible

107 Marx, *Grundrisse*, e.g. pp. 156–7.

108 Karl Marx and Friedrich Engels, *The Communist Manifesto* (London: Pluto Press, 2008), p. 55.

109 Ibid., p. 39; see also Bensaïd, *Marx for Our Times*, p. 122; Raymond Williams, *Keywords: A Vocabulary of Culture and Society* (London: Fontana Press, 1983), pp. 60–9; Crompton, *Class and Stratification*, pp. 28–33; Durou, 'Class Is What', p. 219.

without a ruling class producing them. As individual capitalists, they may bet on the wrong horse, end up in a geographical zone that is losing ground or make poor judgements. Individual capitalists go bankrupt, and tens, hundreds or perhaps thousands of employees may lose their jobs as a result. But only as a class can they create crises that send millions of workers into unemployment in almost no time.

With ecological crises, individual capitalists can destroy wilderness, pollute a city or kill a river. Full-blown ecological crises at the levels we have become accustomed to are only possible because the destruction of nature is part of a global system – capitalism. No capitalist can single-handedly create a metabolic rift. It is only as a class that the ruling class can change the climate, produce the first mass extinction since dinosaurs walked the earth, or produce zoonoses and pandemics at an ever-increasing pace.

It is only as a class that capitalists create a world based on exchange-value and abstract space, dependent on a rift in the social metabolism between man and nature. This does not happen primarily because personal fortunes are astronomical – it is possible only because we have a class that dominates both humans and nature.

The capitalist class creates a world in its own image through creating crises. This the capitalist class cannot do single-handedly. It is not *that* powerful. This happens within certain contexts, frameworks and institutions that also need to be examined. The social power of the capitalist class always exists in interaction with other social relations. In the next chapter we will look at how this happens.

5

Creating Crises in Its Image

The world's best book title continues to be *What Does the Ruling Class Do When It Rules?* The answer Therborn gave to his own 1978 title is: 'It reproduces the economic, political and ideological relations of its domination.'[1] Critical theorists have discussed how this is done by pointing to ideology, violence and economic power, the role of the state and social reproduction, or the production of space, racism, imperialism or natural degradation.[2] How do crises fit in here? Or, perhaps, rather: Do crises fit in here?

If capital accumulation is at the heart of capitalism, what role do economic crises play – given that they are temporary breaches of accumulation? If we see economic crises as mere breakdowns in economic

1 Göran Therborn, *What Does the Ruling Class Do When It Rules?* (London: Verso, 2008), p. 161. It may be added that Therborn's book is not explicitly about crises.

2 On ideology, see Louis Althusser, *On Ideology* (London: Verso, 2008); on economic power, see Søren Mau, '"The Mute Compulsion of Economic Relations": Towards a Marxist Theory of the Abstract and Impersonal Power of Capital', *Historical Materialism* 29, no. 3 (2021), pp. 3–32; on the state, see Therborn, *What Does the Ruling*; on social reproduction, see Tithi Bhattacharya (ed.), *Social Reproduction Theory: Remapping Class, Recentering Oppression* (London: Pluto Press, 2017); on the production of space, see Henri Lefebvre, *The Production of Space* (Oxford: Basil Blackwell, 1991); Neil Smith, *Uneven Development: Nature, Capital, and the Production of Space* (Oxford: Basil Blackwell, 1991); on racism, see Robert Miles, *Racism* (London: Routledge, 1989); on imperialism see Rosa Luxemburg, *The Complete Works of Rosa Luxemburg*, vol. 2, *Economic Writings 2*, ed. Peter Hudis and Paul Le Blanc (London: Verso, 2015); and on nature degradation, see John Bellamy Foster, Brett Clark and Richard York, *The Ecological Rift: Capitalism's War on the Earth* (New York: Monthly Review Press, 2010).

reproduction, are the latter something that happens *despite* the system's crises? If one wants to pick out quotes from Marx, one can see that '*continuity* is the characteristic mark of capitalist production', but is discontinuity then the opposite?[3] If profit is at the heart of the system, what happens to the system when profits evaporate? We can see similar arguments with ecological crises. If capitalism depends on nature to reproduce, is systematic destruction of nature a threat to the reproduction of capitalism? Based on a narrow logic, we may come to the wrong conclusion: that crises must be avoided for capitalism to reproduce. Our analysis has already shown that the opposite is in fact true. The ruling class rules by creating crises in its image. *Crisis is what the ruling class creates when it rules.*

If war, in the words of Carl von Clausewitz, is a continuation of politics by other means, crisis is politics under extraordinary conditions. We have previously discussed how capitalism produces crisis, and we have initiated the discussions on how crisis reproduces capitalism by looking at 'solutions' to the crises, and their similar class character. Now we take this one step further. A full review of how capitalism and crisis reproduce each other would necessarily be a review of the entire political economy and almost every aspect of our societies. Here, we will only make various dips into key political, social and institutional aspects that are central for a ruling class to rule through crises: the state, the historical geography, nationalism, racism, fascism and war. We begin with a phenomenon directly embedded in the nature of crises: the crisis as a state of shock.

Ruling through Discipline and Shock

Crises discipline. The disciplining of the working class goes deeper than the crises, but the crises give the disciplining an extra strength; the whip gains extra force.[4] Capitalists can live longer without workers than workers can without capitalists. This becomes especially clear when the crises strike. Crises cause workers all over the world, to varying degrees, to suffer from increased poverty and unemployment, poorer housing,

3 Karl Marx, *Capital*, vol. 2 (London: Penguin, 1978), p. 182, italics added.

4 On how debt can discipline, see, e.g., David McNally, *Global Slump: The Economics and Politics of Crisis and Resistance* (Oakland: PM Press, 2011), ch. 5.

and even shortages of food and water. This can lead to people taking to the streets and demanding progressive political and economic reforms. We all hope that people rise up and rebel when capitalism no longer provides jobs, housing or food. More often, it is the case that crisis makes workers *more dependent* on their rulers. The crises themselves are good arguments for resistance, but those who have the most reason to protest the crises are often the same people who do not have the room to manoeuvre to resist. Without food on the table, it is difficult to dream of a better world – it is easier to dream of food on the table. For people at risk of losing their jobs, it makes sense to hope that some politician will restore the pre-crisis situation and bring the jobs back. The irony, of course, is that the politicians they are forced to trust are the very ones who caused many of their troubles, and the economy they hope will return is the one that created the crisis in the first place.

Crises are engines for conspiracy theories and disinformation. Crises create shock and chaotic situations that make information dissemination and political communication difficult. One example, excellently described by human geographer Laura Pulido, was when the North American right responded to the massive wildfires on the US West Coast in September 2020 by avoiding the obvious issue – global warming – and instead fuelled massive conspiracies on social media that Antifa was behind the fires.[5] But crises also open opportunities for the ruling class on deeper levels than fake news.

Unemployment, or just the threat of unemployment, can be a more effective way of disciplining the working class than both the police and the military combined. Increased precarisation, poverty and unemployment make it easier to reduce wages, increase exploitation, and thus discipline wage workers. Crises are favourable opportunities for accumulation through dispossession, partly because capital is desperate to expand into new territories, but also because an increasingly precarious situation for workers and the poor across the world makes them even more susceptible to the ruling class order. Economic crises can also be perfect opportunities for capitalists who want to accumulate through the destruction of nature. In economic crises, profits and growth must

5 Laura Pulido, 'Wildfire Rumors and Denial in the Trump Era', in Irma Allen et al. (eds), *Political Ecologies of the Far Right: Fanning the Flames* (Manchester: Manchester University Press, 2024), pp. 57–79.

be restored – suddenly, new oil fields are opened, then environmental standards are lowered, then old environmental plans fall by the wayside.

Precarisation and disciplining come with a timing that further cements class power. In synchronicity, workers' lives worsen and the world descends into shock and panic. As uncertainty spreads, it may seem that there are few options other than to let the ruling class steer the boat (back) to shore, or to let a strong leader save the day. In political theory, the argument can be taken a step further: in the words of the Nazi legal theorist Carl Schmitt, 'sovereign is he who decided on the exception'. The power over the state of emergency gives the rulers not only power in a crisis situation, but even the power to define what is a crisis and what is not a crisis.[6]

After a recession comes an economic upswing, and with new job opportunities emerging, along with ideological apparatuses telling us that we are now out of all misery, it is worth raising the question of whether the cyclical movements in and out of recessions also make resistance more difficult.

With climate change, the state of emergency is arguably not needed, as something less radical can be just as effective. In some kind of well-functioning reason-based society, the climate crisis would obviously have developed into a gigantic crisis of legitimacy. Under capitalism, not only is this not the case, there are even arguments that capitalism can be stabilised both politically and economically due to global warming. The combination of the gravity of the situation and the short time available to solve it might lead many to conclude that we must simply hope for the least bad option to come from above, rather than building the best possible one from below. This becomes particularly relevant when the clearest alternative to neoliberal ecological modernism within the ruling class is far-right climate denial.

Since transformation requires both vast resources and a coordination capacity that only states possess, can we really gamble on solving the climate crisis without relying on those who already have capital or control the states today? The very temporality of the climate crisis makes

6 In relation to the climate crisis, see Joel Wainwright and Geoff Mann, *Climate Leviathan: A Political Theory of Our Planetary Future* (London: Verso, 2020), p. 22; see also Naomi Klein, *The Shock Doctrine: The Rise of Disaster Capitalism* (London: Penguin, 2007), p. 131. For an analysis that takes this a few steps too far with Covid-19, see Giorgio Agamben, 'The Invention of an Epidemic', *European Journal of Psychoanalysis*, journal-psychoanalysis.eu, 26 February 2020.

the political left a bystander to a process in which the premise of all climate negotiations is how to save the planet within the framework of capitalism. Crises always come with shocks, and this has arguably reached new levels with the climate crisis. The problem here is precisely that the climate crisis is a capitalist *crisis.*

Relative Autonomy of the State

It has become a central feature of economic crises that states step in and save the system. Some socialists seem to be equally surprised every time this happens. This may seem ironic since – at least according to some people's view on liberal theory – the market is supposed to be self-regulating. Others consider this to be a socialist measure, since state money for some is almost synonymous with socialism. This is neither irony nor socialism; it is how capitalist states work.

Crisis policies after both 2008 and 2020 have often been described as socialism for the bankers and capitalism for the rest of us, which is what we have heard throughout the era of neoliberalism: risks are socialised and rewards are privatised; private capitalists take the profits when things are going well and leave taxpayers to pick up the bill when things are going badly.[7] This phenomenon is not entirely new. When Hamburg was in deep crisis in 1857 and the city tried to stem the crisis by socialising the capitalists' losses, Marx wrote that 'this sort of communism, where the mutuality is all on one side, seems rather attractive to the European capitalists'.[8]

The state has always played a role in crises of capitalism, but how and to what extent have changed throughout history. In the very early days of capitalism, it was often banks and finance barons who needed to rescue states from bankruptcy (not least because wars were fought on credit).[9]

7 See, e.g., Yanis Varoufakis, 'Crashed: How a Decade of Financial Crises Changed the World: Review', *Guardian*, 12 August 2018; Mariana Mazzucato, *The Entrepreneurial State: Debunking Public vs. Private Sector Myths* (London: Anthem Press, 2013), ch. 9.

8 Karl Marx, *Critique of the Gotha Programme*, in *Karl Marx and Frederick Engels: Selected Works*, vol. 3 (Moscow: Progress Publishers, 1970 [1875]), pp. 13–30.

9 Lars Magnusson, *Finanskrascher: från kapitalismens födelse till Lehman Brothers* (Stockholm: Natur & Kultur, 2020), p. 51; Carmen Reinhart and Kenneth Rogoff, *This Time Is Different: Eight Centuries of Financial Folly* (Princeton, NJ: Princeton University Press, 2009), p. 87.

States have long needed to show a capacity for action when people risked losing their money when banks went bust.

During the nineteenth century, states became increasingly aware of the importance of pursuing an active crisis policy, often stepping in to rescue companies and banks.[10] Perhaps the most distinctive form of state crisis policy in the nineteenth century was the deliberate policy of opening new markets, from China and Egypt to Latin America – creative destruction as imperialist violence. Since the 1930s, as discussed in chapter 1, it has been taken for granted that states in crises guarantee the existence of capitalism. State crisis policy has become an integral component of our entire political economy. This is partly because the ruling class does not dare to wager that the crises always will resolve themselves, partly because politicians in liberal democracies need support from not only (parts of) capital but also (potential) voters who stand to lose a lot from the crises.

There is an abyss between, on the one hand, much liberal rhetoric that describes state interventions as unwanted and too much state regulation as the cause of crisis, and, on the other hand, liberal praxis where states will always intervene to save capitalism under crisis. That capitalism will always need various state regulations also means that there will always be state regulations liberals can blame for this or that crisis. The crux of liberalism is that continuous and pervasive state intervention is crucial to ensure the 'freedom' of the markets, while liberals pretend that it does not interfere.

The crises in Asia in the 1990s furnish an interesting example, because at first glance they appear to be exceptions to the rule of active states as guarantors of capitalism. Just a few years after Mexico was 'saved' by new loans and the Nordic states 'saved' their financial systems, the response of the financial establishment to the crises in Asia sounded very different: let it burn. According to a Morgan Stanley strategist, in the midst of the crisis, 'more bad news' from Asia was needed: 'Bad news is needed to keep stimulating the adjustment process.'[11] President Bill Clinton followed Wall Street and, suddenly, it seemed that Washington would return to laissez-faire and let the crisis resolve itself. Central to the story is that the Asian success stories – such as Malaysia,

10 For Sweden 1857, see Magnusson, *Finanskrascher*, p. 122.
11 Quoted in Klein, *The Shock Doctrine*, p. 267.

South Korea and Thailand – were built on protectionist policies, with foreign investment often excluded and energy and transport in public hands. The Asian countries would not be helped by laissez-faire; they would be broken by realpolitik.

Capitalist states must save capitalism, but this can be done in very different ways. The state – this large and somewhat elusive institutional ensemble – is one of many institutions in our society, but also quite special because it has an overall responsibility for maintaining some kind of cohesion in the social formation of which it is part.[12] The state comes in different varieties, but its overall responsibility for holding both society and capitalism together becomes clear in times of crisis. States always reproduce the capitalist class through crisis, and we can further explore this through two different approaches: either through actors and individuals, or through structures. We will briefly discuss both.[13]

People at the top of the state apparatus usually come from the upper strata of society themselves. They may have such high salaries that they identify with those who accumulate capital or such good incomes that they can invest consciously on the stock market. We also know that people with money and power seek out others with power. The capitalist class is the class with money and can buy – directly or indirectly – both friends and influence. Politicians and high officials also have an interest in networking with those with money 'to get things done'. For some, government and business are overlapping labour markets: senior politicians can take top executive jobs in business and vice versa (as when Rex Tillerson, CEO of the world's largest non-state oil company, slid straight into Donald Trump's cabinet, or when former politicians are constantly recruited as consultants and executives in oil companies in Norway). In countries like the US, politicians can be directly indebted to capitalists who have paid for their election campaigns, and even in Sweden, direct favours between

12 See, e.g., Bob Jessop, *State Power: A Strategic-Relational Approach* (Cambridge: Polity Press, 2008), p. 79; Therborn, *What Does the Ruling*, pp. 129–244; Stanley Aronowitz and Peter Bratsis (eds), *Paradigm Lost: State Theory Reconsidered* (Minneapolis: University of Minnesota Press, 2006).

13 See, e.g., Nicos Poulantzas, 'The Capitalist State: A Reply to Miliband and Laclau', *New Left Review* 95, no. 1 (1976), pp. 63–83; Nicos Poulantzas, *State, Power, Socialism* (London: Verso, 2000); contra Ralph Miliband, 'The Capitalist State: Reply to Nicos Poulantzas', *New Left Review* 59 (1970), pp. 53–60; Ralph Miliband, *The State in Capitalist Society: An Analysis of the Western System of Power* (London: Quartet Books, 1973).

capitalists and politicians are not unknown phenomena – a fact that is glaringly visible in the privatisation of health care in Stockholm.[14]

Who sits where in the state apparatus matters a lot, but this tells us nothing about why *every* state in every crisis must save capitalism. Even in countries where parties explicitly based on struggle against the ruling class have ruled for decades, as in post-war Scandinavia, the ruling class has been able to reproduce capitalism relatively unproblematically. Why? One can point to individual explanations here as well, such as the fact that the Norwegian Social Democrats were extremely anti-communist and took Norway into NATO, or that there were close ties between Swedish Social Democrats and some key capitalist families. To understand why this seems to happen necessarily in all capitalist countries, we must also analyse the structural relations between the capitalist class and the state.

Capital accumulation is not only profit for owners, exploitation of workers and destruction of nature; it also leads to wages, taxes and growth on which states depend to maintain basic functions such as police, military and bureaucracy – and also to distribute welfare. The capitalist class can exert great influence over states by threatening to move if they do not obey, or by buying and selling government bonds. At the same time, the capitalist class depends on laws, regulations, systems, available labour and often a certain predictability that only the state can guarantee. (We will not discuss grey areas here, such as criminal gangs offering similar protection.) From a structural perspective, states are capitalist because the capitalist class and political power are mutually dependent on each other for the functioning of the ruling order. And the ruling order must be maintained in order for both the capitalist class and the existing states to reproduce themselves. Economic crises then become events that must generally be resolved – or used – in order for both capitalism and the modern state apparatus to function at all.

When economic crisis ends up directly on the table of states, both structural and personal links between state and capital might be exposed. Through crisis management, states become actors contributing to both creative destruction and various forms of spatial and temporary fixes.[15]

---

14 On Sweden, see, e.g., Göran Therborn, *Kapitalet, överheten och alla vi andra* (Lund: Arkiv förlag, 2018), pp. 107–18; Andreas Cervenka, *Vad gör en bank?* (Stockholm: Natur & Kultur, 2017), pp. 38, 111, 138.

15 Jessop, *State Power*, p. 11.

States are central to the daily reproduction of the class society that creates economic crises, from the legalisation of exploitation (private property is protected in constitutions) to facilitating accumulation through dispossession. For ecological crisis, the state's legitimisation of an exchange-value relationship with nature lays the groundwork for accumulation through natural destruction and metabolic rifts.

Can one imagine that the class character of the state changes if, for example, it becomes less attractive for rich people to work in that particular state? Martin Gustavsson and Andreas Melldahl have shown that the share of the 'owning class' working in the Swedish state apparatus fell dramatically when the Social Democrats came to power in 1932, and the number working in public administration continued to fall during neoliberalism, as it has become less attractive for wealthy people to work in the Swedish public sector.[16] Does the state become *less* capitalist when fewer people from the upper strata of society want to work there? No. Here, structures trump persons. If anything, there was the opposite development during the twentieth century: while there is no reason to believe that it became more common for people with affluent backgrounds to work in the public sector during neoliberalism, states became arguably *increasingly* trustees of the common interests of the capitalist class. In crises today, there is absolutely no doubt: states are willing to go to extreme lengths to save the system because for them it is so much about saving themselves.

Since the rational decisions of individual capitalists can lead to irrational consequences, it becomes the historical task of states to intervene in this anarchy and bring the system together. Therefore, it is in the interest of states to create general rules and laws that will promote growth, to create a 'good business climate'. Or, according to Marx and Engels's famous formulation in *The Communist Manifesto*: the 'executive of the modern state is but a committee for managing the common affairs of the whole bourgeoisie'.[17]

Crises can give new vitality to capitalism because there is a *relative autonomy* between the state and class relations, between political and economic power. In contrast to how ruling classes in previous societies

16 Martin Gustavsson and Andreas Melldahl, 'The Social History of a Capitalist Class: Wealth Holders in Stockholm, 1914–2006', in Olav Korsnes et al. (eds), *New Directions in Elite Studies* (New York: Routledge, 2018), pp. 187, 189–91.

17 Karl Marx and Friedrich Engels, *The Communist Manifesto* (London: Pluto Press, 2008), p. 36.

generally controlled both the state and the economy, today the capitalist class does *not* directly possess political power.[18] Nicos Poulantzas argued that the capitalist state is neither a thing, object or instrument that any class or group can fully control at its will (i.e., no autonomy), but nor is it a freely independent subject that can do whatever it wants (i.e. full autonomy). The state is, rather, a *material condensation of social relations*. The state is an expression of – and part of – the social relations that exist in society at large.[19] This relative autonomy gives crises both a vent and an openness, while setting clear boundaries. It is a key to understanding how crises can open up political discussion while reinforcing and legitimising the system. To exemplify, we can take a short detour via the Soviet Union.

Elmar Altvater published an interesting book on crises in 1991, in which he could not avoid dealing with the fact that 'half the world' had just collapsed in a gigantic crisis. In the Soviet political economy, according to Altvater, there was no relative autonomy between the economy and politics, because production was governed by political and not economic goals. While Eastern bloc countries were based on the 'primacy of politics', market economies are based on the 'primacy of the economy'. Minor crises in the East could therefore be averted and hidden under politics, but because there was no relative autonomy, all major economic problems were directly transformed into political crises. Crises that could not be concealed or stopped escalated into crises of legitimacy for the entire social formation.[20] According to Altvater, the so-called socialist states did not collapse because of limitations in material welfare, but rather because 'the institutions of society were insufficiently flexible in adjusting to crisis tendencies that had been concealed

---

18 On relative autonomy, see Nicos Poulantzas, *Political Power and Social Classes* (London: New Left Books, 1973); Poulantzas, 'The Capitalist State'; Poulantzas, *State, Power, Socialism*. On economic and political power see, e.g., Ellen Meiksins Wood, *The Origin of Capitalism: A Longer View* (London: Verso, 2002); Mau, 'The Mute'; and the whole of *Historical Materialism* 29, no. 3 (October 2021). In relation to hegemonic crisis, see Stuart Hall et al., *Policing the Crisis: Mugging, the State and Law and Order* (London: Macmillan, 1978), p. 201.

19 Poulantzas, 'The Capitalist State', p. 74; Poulantzas, *State, Power, Socialism*, pp. 73, 123, 128–39.

20 Elmar Altvater, *The Future of the Market: An Essay on the Regulation of Money and Nature after the Collapse of 'Actually Existing Socialism'* (London: Verso, 1993), pp. 22, 47. Originally published as *Die Zukunft des Marktes* (Münster: Westfälisches Dampfboot, 1991).

for too long'.[21] We will not go further into the political economy of the Soviet Union here, but Altvater's analysis gives us a very good starting point for understanding capitalism.

In capitalist societies, the general rule is *not* that economic or ecological crises automatically become political crises. They might develop into political crises, but this rarely happens immediately. Gramsci pointed this out in 1926, in his discussion of the rise of fascism. Serious economic crises do not have immediate consequences in the political sphere because the state apparatus is much more resistant than often imagined: it manages 'at moments of crisis, in organizing greater forces loyal to the regime than the depth of the crisis might lead one to suppose'.[22]

One reason for this is precisely the institutional, social and ideological distinctions – porous, but existing – that reside between politics and economics. If there were no autonomy at all, economic and ecological crises would immediately become political; if political autonomy were total, the state could let capitalism fall and then continue to be the same state in a new economy.

This is where Jürgen Habermas goes wrong. His *Legitimation Crisis* was originally published in German in 1973, and the year is important, because Habermas's starting point was to ask why there were no economic crises even though there were contradictions in capitalism that should create such crises. (This was a relatively common approach to crises, up until 1973.)[23] Habermas's answer was that, because of the relative separation of economic and political power under capitalism, economic crises can be diverted and directly become other types of crises: political, administrative, social and so on. Today, fifty years later, we can see how Habermas was caught up in his time. The analysis may explain 1968, but, on the other side of neoliberalism – a sea of crises later – we have to conclude that Habermas was wrong. Contradictions in capitalism certainly create economic crises, and the relative autonomy of the state should rather be understood as the reason why these do not directly and necessarily become social or political crises.

---

21 Ibid., p. 23.

22 Antonio Gramsci, 'The International Situation and the Struggle against Fascism, 1926', in David Beetham (ed.), *Marxists in the Face of Fascism* (Chicago: Haymarket, 2019 [1926]), p. 125.

23 Jürgen Habermas, *Legitimation Crisis* (London: Polity Press, 1976).

The state certainly does not manage the common affairs of the whole bourgeoisie, if by this we mean that it benefits all capitalists, all the time.[24] If the state cannot help everyone in a crisis, which fractions should benefit and which should be sacrificed? Even fiscal and monetary policy, which is sold as the common good of all, always favours some at the expense of others. The same goes for technological change: Which technology does the state choose to invest in and who benefits from it? State crisis policies also strike as watersheds in regional development: Which regions will be sacrificed in the name of progress and which cities will become the power centres of tomorrow? All this is happening at the same time as states and nations seek to strengthen their positions vis-à-vis others. Even when politicians know who they want to benefit – what is the best policy to do so? The crisis policy closest at hand is always one of yesterday's policies, but relying on yesterday's (hegemonic) economic policies can create serious challenges for tomorrow, as with the attempt to solve the 1929 crisis with laissez-faire, the 1970s crisis with Keynesian policies, or the 2008 crisis with lower interest rates and increased debt.

One problem is that states must bail out the capitalist class in crises, but, if the ruling class knows this, it can lead to further risk-taking and crazy speculation – which can be like pouring gasoline on the fire. Charles Kindleberger has argued that governments should be lenders of last resort to save the system, but, at the same time, *pretend* they are not, so that companies do not take too much risk.[25] This is, in truth, a desperate realpolitik attempt to find a fruitful liberal crisis management.

This relative autonomy allows politicians to place themselves both inside and outside crises. On the one hand, politicians can argue they had nothing to do with the causes of crises; on the other hand, politicians see crises as challenges for them to solve. Even in a bourgeois democracy, various environmental problems can be hidden and economic figures manipulated to try to avoid crises, but precisely because of their relative autonomy, politicians can also welcome the crises. While blaming others for creating the crises, politicians can ride on episodes of shock and uncertainty to win elections by presenting

24 See Stuart Hall et al., *Policing the Crisis*, pp. 201–8.

25 Charles P. Kindleberger, *Manias, Panics, and Crashes: A History of Financial Crises* (New York: John Wiley and Sons, Inc., 2000), ch. 10.

themselves as strong leaders in troubled times. We know from history that this can work both ways. In part because of their own policies and in part because of changes outside the domain of politics, a politician like Herbert Hoover could lose out in crises, while Franklin D. Roosevelt and Margaret Thatcher could emerge as winners.

Relative autonomy even provides some openings for change from below. Social relations are expressed differently in/through the state and politics, making states sites of political, economic and social struggles. The left and the labour movement must do everything they can to avoid the worst consequences of any crises. But we need to be aware that this happens within the framework of capitalist states, where states have to save the capitalist class in order for it to continue to accumulate capital (structurally), and where there are close links between capitalists, top politicians and senior officials (personally). And we must add that many politicians genuinely believe that what is good for business is always good for all of us (ideologically). Crises are periods when class struggle must necessarily be waged on various fronts – there is no choice – but when the conditions for thinking beyond capitalism seem even worse than in other circumstances.

When the organised working class is at its strongest and the capitalist class is relatively weak – and when the sun is at its zenith, and there is perhaps a frightening dictatorship in the East – a compromise between the classes is the best that can be hoped for within the framework of the capitalist state. Relative autonomy even allows sections of the capitalist class to see trade unions and social reform as acceptable solutions to crises. There have been crises where parts of the capitalist class reasoned that it was better to let a non-revolutionary social democratic party and trade unions discipline the working class than to let the fascists do it.[26] According to Therborn, certain interventions may 'well go against prevailing ruling-class opinion', while 'objectively furthering or maintaining its mode of exploitation and domination'.[27] Roosevelt's New Deal faced sharp criticism from the capitalist class, who argued that the president had betrayed his class. Commenting on the growing opposition to the New

26 For Norway, see Harald Berntsen, 'Kapitalismens kriser og arbeiderbevegelsens svar', *Sosialistisk Framtid*, 1 March 2021.

27 Therborn, *What Does the Ruling*, p. 148.

Deal in business circles in the spring of 1935, Roosevelt commented that it was rather his class that had betrayed him.[28]

The organisational framework of the state is a precondition for discursively turning crises into situations where we all have common interests. The state is an arena for class struggle, but it is anything but a neutral playing field. Yet it must be presented precisely as objective and impartial. The state is an expression of class antagonism, but is often presented as a mediator between classes.[29] The *appearance* of the capitalist state as neutral is not based on nothing; our capitalist states and their bureaucracies have a far more impersonal form compared to pre-capitalist states. Where kings throughout history could point to God for legitimacy, modern capitalist states often derive legitimacy by acting in the common good of the nation or the people. Perhaps the most effective way to conceal class power, and thus to create a world in its image, is to make a specific class interest appear to be the interest of all. To understand how this works, we need to talk nations and hegemony, and along the way we must discuss the historical geography of crises.

Thoughts on the Similar but Different Historical Geography of Crises

Capitalism and its modern crises have been walking together on the earth for about 200 years, and this journey can also be read as a specific historical geography. It is tempting to sum up the journey with this general statement: economic and ecological crises have moved over time from being local and regional phenomena to becoming national and international, and finally global. But, as soon as we have formulated this claim, we have to make it more complex. It is complicated by (at least) four conditions.

First, there are differences between ecological and economic crises, to which we shall return. Second, the journey has certainly not been without setbacks and key exceptions. International connections existed before the nineteenth century; cross-border flows of goods and services (relative to global production) peaked in the 1870s and only returned to

28 See ibid.; see also Steve Fraser, 'American Labour and the Great Depression', *International Journal of Labour Research* 2, no. 1 (2010), p. 13.

29 Stuart Hall et al., *Policing the Crisis*, ch. 7.

that peak in 1970; half the world lived in so-called socialist states for much of the twentieth century; and the US even lost imperialist wars.[30] Third, the concept of the 'global' has different meanings. Since crises are embedded in the very DNA of capitalism, the 'global' – or at least the global potential – was arguably present from day one. According to Marx, the tendency to create a world market is immediately given in the very concept of capital, and, according to Luxemburg, it is in the nature of capitalism that it needs to have all the territories of the world at its disposal.[31] And, fourth, although crises today are global, they still take highly local and regional forms. Despite exceptions and empirical and theoretical difficulties, it is nevertheless interesting to discuss the common historical geography of economic and ecological crises.

In sharp contrast to what would come later, early speculative bubbles had very little geographical spread. The Tulip Crash of 1636–37 remained a relatively local phenomenon and the Mississippi Bubble and the South Sea Bubble of 1720 – despite being part of the violent rise of capitalism through colonialism, trade and general plundering of people – were nevertheless geographically confined to a few central locations. Capitalism still lacked a global infrastructure of production chains, finance and trade networks.

The claim that crises have moved from being more local to fully global phenomena may sound abstract, but it is in fact very concrete. We see that crisis as an economic term became common in England in the eighteenth century, while it was established in German only in the nineteenth century, which to some extent reflected the actual historical geography of economic crises. German correspondents in England warned their compatriots in 1825 of the 'imminent crisis' that might be coming, and by the following year 'crisis' was used to describe economic problems and bankruptcies in Germany as well.[32] A report by the Cologne Chamber of Commerce in 1837 stated that 'because in the last two decades our province had entered into significant direct and indirect relations with North

30 Matthew C. Klein and Michael Pettis, *Trade Wars Are Class Wars: How Rising Inequality Distorts the Global Economy and Threatens International Peace* (New Haven, CT: Yale University Press, 2020), p. 23.

31 Marx, *Grundrisse*, p. 408; Luxemburg, *The Complete Works*, p. 261; see also McNally, *Global Slump*, p. 37.

32 Reinhart Koselleck, 'Crisis', *Journal of the History of Ideas* 67, no. 2 (2006), pp. 366, 389.

America, it was inevitable that the adverse effects of this crisis would be felt by our commerce and factories'.[33] Capitalists in Germany discussed in real time how the crises crossed national borders.

Concerning ecological crisis, the hypothesis of local origin is surely complicated if we go back to the sixteenth, seventeenth and eighteenth centuries when Europeans, in addition to spreading death through massacres, wars and displacement, also exported diseases and epidemics as well as major ecological problems through deforestation and the rapid spread of European crops, weeds/grasses and animals.[34] With metabolic rifts, the 'local' origin is easier to pinpoint. In the heartlands of capitalism, rural soils were depleted in the nineteenth century, and in cities, nutrient-rich faeces became a local environmental problem instead of being returned to the soil. Within cities, environmental problems were initially confined to individual neighbourhoods, but became city-wide problems as diseases spread through water.[35] When nascent capitalist agriculture needed fertiliser, first bones from the Napoleonic battlefields were looted to spread across fields, and then phosphate- and nitrogen-rich bird droppings, guano, began to be imported from Peru and Chile to Europe.[36] The first shipment of guano arrived in Liverpool in 1837 and by the 1860s the Peruvian Pacific coast had been completely stripped of its nutrient-rich gold. In addition to direct ecological degradation in Chile and Peru, the 'trade' involved thousands of Chinese workers who were shipped to Peru in slave-like conditions. The early internationalisation of the metabolic rift also included a veritable war between poor countries (Peru and Bolivia against Chile) over guano resources, the demand for which was controlled by the core capitalist countries. US capitalists annexed ninety-four islands, large and small, between 1856 and 1903 (and nine are still US property) in their chase for guano, not least because slave plantations producing tobacco saw

---

33 Ibid., p. 391.

34 See, e.g., Sverker Sörlin, *Antropocen: en essä om människans tidsålder* (Stockholm: Weyler förlag, 2017), pp. 142–3.

35 See also Ian Angus, *Facing the Anthropocene: Fossil Capitalism and the Crisis of the Earth System* (New York: Monthly Review Press, 2016), p. 122; Karl Marx, *Capital*, vol. 3 (London: Penguin Classics, 1981), p. 195.

36 John Bellamy Foster, *Marx's Ecology: Materialism and Nature* (New York: NYU Press, 2000), p. 151; John Bellamy Foster and Fred Magdoff, 'Liebig, Marx, and the Depletion of Soil Fertility: Relevance for Today's Agriculture', *Monthly Review* 50, no. 3 (1998), pp. 32–45.

dramatic declines in soil fertility.[37] If nothing else, this shows how even ecology and the imperial, colonial and racist history of capitalism are intertwined. With the industrial production of synthetic fertilisers from 1913, the metabolic rift moved to new levels. Short-term attempts to solve the nutrition issue always led to a new ecological crisis elsewhere and at a higher level.

Fossil capital has its own geographical histories. Historians have long assumed that the shift in England from hydropower to coal in the early nineteenth century was due to coal being cheaper and more stable. As mentioned, Andreas Malm's research has shown that hydropower was both cheaper and more stable, but to understand the transition we then need to analyse a combination of class and geography.[38] Whereas a water mill could only be located in specific places (for example, near a waterfall), a coal plant could in theory be located anywhere. Rather than moving the factory to the energy source and attracting workers there, the factory could be located in towns populated by poor people who had been driven from the countryside. The transition to coal became a cornerstone in the production of global capitalist spaces – that is, abstract spaces. In its perpetual pursuit of profits, and in its need to discipline and control workers, capital can constantly move geographically and create new landscapes.

In 1825, England accounted for 80 per cent of all greenhouse gas emissions from burning fossil fuels. Not long after, the US and Germany caught up, and fossil fuels became a central component of wars and imperialism, as well as fuel for the post-war 'golden age' and cold wars.[39] Today, we have a global economy dependent on oil, gas and coal. Abstract space produced by concrete fuel.

Economic crises of the nineteenth century often travelled by boat and

---

37 Rikard Warlenius, 'Inledning: Fyra debatter och en begravning', in Rikard Warlenius (ed.), *Ecomarxism: Grundtexter* (Stockholm: Tankekraft, 2014), pp. 30–3; Brett Clark and Richard York, 'Rifts and Shifts: Getting to the Root of Environmental Crises', *Monthly Review* 60, no. 6 (November 2008), p. 17; Foster, *Marx's Ecology*, pp. 149–54.

38 Andreas Malm, *Fossil Capital: The Rise of Steam Power and the Roots of Global Warming* (London: Verso, 2016).

39 Andreas Malm, 'Long Waves of Fossil Development: Periodizing Energy and Capital', *Mediations* 32, no. 1 (2018), pp. 17–40; Angus, *Facing the Anthropocene*, ch. 8; on fossil fuels and political struggle/class struggle, see Timothy Mitchell, *Carbon Democracy: Political Power in the Age of Oil* (London: Verso, 2013).

train, running on coal, often between the core countries of capitalism, not infrequently via their colonies. According to Luxemburg, the development of the railway network more or less reflected the penetration of capital: 'In Europe, the railway network grew most rapidly in the 1840s, in the US in the 1850s, in Asia in the 1860s, in Australia in the 1870s and 1880s, and in Africa in the 1890s.'[40] And, with the railways came capital, credit, debt – and crises.

The economic crisis of 1857 has been called the first international recession.[41] A German consul in the United States was able to report back to Berlin that 'attempts to identify the origin of this crisis have shown that it is everywhere and nowhere'. The crisis was perceived by many as a world crisis because it was less geographically confined to specific countries, and it became interesting to talk about a 'history of the world economy'.[42] But still the more serious consequences of the crisis were limited to a handful of countries and stock markets. It is more common to call the 1873 crisis the first truly international crisis of capitalism, and it has even been called the 'first synchronized global financial crisis'.[43]

The Great Depression is usually called a world crisis because the world capitalist system was considered fully developed.[44] But, here, the Soviet Union is a massive exception. Although still quite isolated and weak, the Soviets were experiencing relatively good economic development (albeit from a very low base) when the West entered the Great Depression in 1929. While the crisis of the 1930s spread back and forth between countries and continents, the crisis of the 1970s, according to Mandel, was the first to hit all the major imperialist powers simultaneously.[45] The rapid rise in oil prices came from the 'periphery' of capitalism, but struck directly at the whole machinery. The crisis did not initially affect the so-called socialist countries, which rather experienced

40 Luxemburg, *The Complete Works*, p. 304.

41 Michael Roberts, *The Long Depression: How It Happened, Why It Happened, and What Happens Next* (Chicago: Haymarket, 2016), pp. 31, 302.

42 Koselleck, 'Crisis', p. 392.

43 Klein and Pettis, *Trade Wars Are Class Wars*, p. 8.

44 See Rikard Štajner, *Crisis: Anatomy of Contemporary Crises and (a) Theory of Crises in the Neo-imperialist Stage of Capitalism* (Belgrade: KOMUNIST, 1976), p. 165.

45 Ernest Mandel, *The Second Slump: A Marxist Analysis of Recession in the Seventies* (London: NLB, 1978), p. 9.

good years when the capitalist economy was in crisis.[46] But, after a while, these countries also faced some difficulties because they were linked to the West by trade. Nevertheless, Štajner argued, this was not 'their crisis', concluding that there was no world crisis, but the whole world had to pay the bill for the crises of capitalism.[47]

It is really since the disappearance of the Soviet Union and the incorporation of China into world capitalism that we can talk about truly *global* economic crises, understood as events that affect more or less the whole world within a relatively short period of time. However, the first crises after the fall of the Berlin Wall remained regional in nature, as the capitalist core was not affected by the problems of the 1990s in Latin America, Africa or Asia, but was able to use them to its advantage.

The 2008 crisis was different. Now production, trade and not least financial systems were highly internationalised and the crisis was truly global. Of the 104 countries for which the World Trade Organization (WTO) collects statistics, all experienced a fall in both imports and exports between the second half of 2008 and the first half of 2009. All countries and all types of traded goods – without exception – experienced a decline.[48]

But records are made to be broken. The economic contraction in 2020 was, according to the IMF, unprecedented in its speed and synchronised nature.[49] Never in modern capitalism have we seen an economic crisis hit so suddenly and so simultaneously on a global scale. Never has economic activity contracted so extensively across the globe in such a short period of time. Never before have there been similar decisions to shut down so much of the economy, and the result was that global GDP fell 20 per cent from early 2020 to 10 April.[50]

The first time people became aware that an ecological crisis could be global, according to Ian Angus, was when it was realised that freons

46 Ibid., pp. 147–56.

47 Štajner, *Crisis*, p. 172; see also Rick Kuhn, 'Henryk Grossman and the Recovery of Marxism', *Historical Materialism* 13, no. 3 (2005), pp. 90–2, versus Mandel, *The Second Slump*, pp. 147–56.

48 ILO, *Global Employment Trends 2011: The Challenge of a Jobs Recovery* (Geneva: International Labour Office, 2011), p. 3; Adam Tooze, *Crashed: How a Decade of Financial Crises Changed the World* (New York: Viking, 2018), pp. 80, 159, 161, 163.

49 IMF, *World Economic Outlook: Managing Divergent Recoveries* (Washington, DC: IMF, April 2021), p. xvi.

50 Adam Tooze, *Shutdown: How Covid Shook the World's Economy* (London: Allen Lane, 2021), pp. 5, 107.

damage the ozone layer.[51] Today, it goes without saying that the climate crisis is global, but it is global in a different way from the economic crises of 2008 and 2020. Whereas economic crises are global in the sense that they affect people (almost) everywhere, today's climate crisis is planetary in the sense that the entire globe is affected, without a single geographical exception. Here, we also have to make a distinction between climate change and pandemics as ecological crises. While the climate crisis is global because it involves the atmosphere, Covid-19 had a global geography similar to the economic crises, as it always travelled alongside people and their economic and social activities.

The crises of capitalism are not only *becoming* more global – they have contributed to capitalist globalisation. The temporary solutions seem to elevate the problems to a higher and more serious level. Early problems of capitalist agriculture were solved first geographically with the pursuit of natural fertiliser (and imperialism and war), and then with a technological fix (and artificial fertiliser). More chemical toxins and increased use of oil in the agricultural process led, in turn, to polluted water, dead seas and global warming. The switch from wood to coal as an energy source solved local problems of deforestation but created global warming; the problem of horse excrement on the streets disappeared when cars came along; today, many will meet global warming with geoengineering (which takes the metabolic rift to a new level) or electrification of society (which requires such quantities of energy and metals like lithium and cobalt that it will necessarily lead to new ecological crises).[52]

Each economic crisis is followed by a new exaggerated optimism about the future, and new booms take the economy to new levels. Out of the creative destruction come spatial and temporal fixes that reorganise capitalism and that, in turn, speed up circulation.[53] Crises are decisive moments in the creation of extended reproduction. At the beginning of each recovery, the economy expands as new markets appear, stimulating the activity of the means of production. Such new markets may result

51 Angus, *Facing the Anthropocene*, p. 123.

52 See Clark and York, 'Rifts and Shifts', pp. 17–22.

53 See, e.g., David Harvey, *The Limits to Capital* (London: Verso, 1999); Jamie Gough, *Work, Locality and the Rhythms of Capital: The Labour Process Reconsidered* (London: Psychology Press, 2003); Noel Castree and Derek Gregory (eds), *David Harvey: A Critical Reader* (Malden, MA: Blackwell, 2006).

from the geographical expansion of capitalist production, the emergence of new sectors of production and technological advances, the privatisation of public services, geopolitical and geographical changes, sudden changes in the relationship between competitors, or war or state-sponsored armaments. Capitalism gets 'back on track', at higher levels.[54]

We have already seen how our ecological and economic crises have the same class character and are rooted in the same system driven by the profit motive and based on the distinction between exchange-value and use-value. Now we have also seen how the crises have a similar, but not identical, historical-geographical development: this is another argument that we are dealing with two *different*, but *capitalist*, crises. For those who refuse to acknowledge that economic and ecological crises are rooted in capitalism, the common historical geography of the crises must seem like an eternal puzzle.

Global Questions, National Answers

The globalisation of capitalism does not break down any borders, it constantly creates new ones. Walls are not being torn down, they are moved and rebuilt. In previous chapters, we saw how some contradictions were central to the reproduction of capitalism through crises: an abstract world is built on principles of exchange-value that always clash with concrete use-values; crises both shake and stabilise capitalism; a series of 'rational' decisions lead to 'irrational' crises. Here comes another: although capitalism and its crises have become increasingly global, the political and social responses remain highly national, nationalistic and racist.[55]

The crises of capitalism legitimise states and nations. The state becomes more often than not the organisation that will solve the crises,

54 For an overview of how capitalism has spread geographically through seventeen different cycles from the 1816–25 cycle to the 1953–58 cycle, see Ernest Mandel, *Late Capitalism* (London: Verso, 1978), p. 110. For more on spatiality and temporality in relation to crisis, see also David Harvey, *The Condition of Postmodernity* (Oxford: Blackwell, 1990), p. 232; Bob Jessop, 'The Symptomatology of Crises, Reading Crises and Learning from Them: Some Critical Realist Reflections', *Journal of Critical Realism* 14, no. 3 (2015), p. 246.

55 See also Michael Löwy, 'Daniel Bensaïd: A Marxism of Bifurcation', *International Viewpoint*, 28 June 2020.

and the nation an imagined community that is needed to make this happen. In fair weather, global political solutions may appear to be underway, but as soon as the wind starts to blow, both politicians and the ruling class return to their nation-states. Ties may be forged more strongly with military allies, and rhetoric against the alliance's enemies intensifies to rally the nation.

The national 'we' emerges to meet the challenge as soon as crises enter the political agenda. This starts with the explanations of the origins of particular crises. When economic crises are explained solely in terms of international business cycles, there is nothing an individual country like Sweden can do. The crises then become a flood from outside attacking 'us'. And there is a point here. Many crises would have emerged regardless of what, for example, Swedish politicians had done. But the fact that crises can be triggered outside Sweden's borders is not the same as Sweden being an independent entity that happens to be hit by the ravages of the world economy. Here, we must beware of spatial fetishism: seeing only geographical dimensions (like nation states) when, in fact, the issue goes deep down into social relations (like class). Crises spread – as we have seen – according to specific geographical patterns, but, if we focus one-sidedly on this, we risk missing how they arise from trends and relations in the economy itself. The Swedish economy must be seen as an integral part of the global economy; rather than wildfires in Sweden and California being merely different events on different continents, we need to emphasise that they come from the same source. This may seem obvious, but for those reading newspapers or watching TV during a crisis, it is far from the case.

There have, of course, been attempts to implement the liberal dream – a capitalism without crises – even at an international or global level. The original idea of the International Monetary Fund and the World Bank was to stabilise the world economy through regulations (under Keynesianism), and later – ironically – to try to do the same through deregulations (under neoliberalism). When crises come, it quickly becomes apparent that the IMF, World Bank and World Trade Organization (WTO) are composed of (nation) states that want to save themselves. Similarly, the global climate summits consist of (nation) states that meet to discuss solutions to the global climate crisis. The internationalisation of economic life, according to Štajner, gave birth to internationalisation of economic crises, 'but without any barriers and adequate (internationalised) tools for their suppression'.[56]

The responsibility to carry out crisis management almost always falls on the (national) state. This is partly an institutional issue. It is simply where a significant amount of political power resides. In economic crises, this is where monetary and fiscal policy exists and this is where counter-cyclical policy is created. For the climate crisis, the situation is such that no other power structure could orchestrate the transition that is necessary at this point in time. But states are not neutral actors intervening in crises: they are capitalist states that have to deal with the crises of capitalism.

How tensions between global issues and national responses play out varies between crises and between countries. If we keep the focus on the Global North, we see how politicians must take care of the capitalist state they are set to govern; they must save 'their' capitalist class (and thus save the system), and try to win the support of the people at the next election.[57] The state must simultaneously save itself, put the blame somewhere else and (try to) export problems and unemployment to neighbours and competitors. The energy crisis of 2022 was, arguably, caused solely by Vladimir Putin; US politicians used the corona pandemic of 2020 to intensify the conflict with China; the euro crisis a decade earlier was often portrayed as a conflict between Greece and Germany, and in the 1970s it was oil-producing countries in the Middle East that were made responsible. And so on.

The quintessence of a liberal understanding of crises is to reduce all problems to individuals and nations. Class disappears. The claim of the bourgeois state to represent all individuals – equality before the law, universal suffrage, universal human rights – becomes a central component in legitimising the crisis management of states. In times of political and social instability, in uncertain and unpredictable situations, where it seems inevitable that many will be affected and when politicians need to communicate simple messages, an easy solution to hand out is to unite the nation.

When capitalist hegemonies are articulated through national frameworks and soaked in national chauvinism, it is perhaps no wonder that

56 Štajner, *Crisis*, p. 136.

57 Mandel, *The Second Slump*, p. 47.

crisis management resembles national football more than class struggle. The system is set up to create suspicion, states look at each other with scepticism, and everyone wonders how 'we' can get off cheaply. How can 'we' profit from the climate crisis? Can 'our' companies avoid firing people in times of recession? Can 'we' avoid bankruptcy? Will 'we' have lower unemployment than other countries? Or will 'we' perhaps get more vaccines? The fact that 'our' crisis policy may affect workers in other countries becomes secondary.

Times of crisis (unfortunately) rarely mean a boom in international solidarity. Even parts of the left end up in protectionist positions when politics is defined by the borders of the nation state. One key to the problems is the form of crises: the shock and distress demand a solution *here* and *now*. The crisis requires immediate concrete answers to concrete problems. A left that does not understand the nature of the crises – and may go around howling because they think 'crisis' in Chinese means 'opportunity' – can end up wrong-footed.

The social democratic parties moved – in the words of the Norwegian historian Harald Berntsen – during the interwar period from the idea of an international class struggle against capitalism to national class cooperation in the international struggle against competitors.[58] This movement became *Folkhemmet* (People's Home) in Sweden and *Hele folket i arbeide* (Whole People at Work) in Norway. These slogans show the power that the social democrats and the reformist labour movement actually came to have, but also the limits of that power. That the working class was a strong force in creating a hegemonic national 'we' is indeed one of the exceptions of history. (Not all workers were included in the 'we': Sweden, for example, had entry and exit bans for Roma between 1914 and 1954, and Sami and others were forcibly sterilised until the mid-1950s.) The norm in capitalism is that (parts of) the capitalist class forms the core of the prevailing hegemony, without much interference from the organised working class. The hegemonic power of a class to *represent the nation* is the prerogative of the capitalist class. The illusion of a common interest is central to how we understand hegemonic power: to simultaneously dominate other classes and groups in society *and* present one's particular class interests as the common interests of society or the nation. To do this, the ruling class must extend its authority

58 Berntsen, 'Kapitalismens kriser'.

beyond merely owning the means of production, and into spheres such as civil society and the state.[59]

If the capitalist class constitutes only 1 to 2 per cent of the population, and in itself contains internal tensions, can such a small number of people carry a hegemony alone? In principle, yes. This is more about social power, ownership and central control functions, and less about the number of people involved. But, in practice, no. First, we can start by including large sections of the petty bourgeoisie – which is around ten times larger than the capitalist class – in this hegemonic bloc. Then, broader hegemonic blocs are always created through various alliances, from more long-term social alliances – as when reactionary religious or social forces ally with a right wing over time – to classical alliances between the capitalist class and what is sometimes called the labour aristocracy, or with large and even small peasants, or even the lumpenproletariat.[60] A further analysis of this would require that we discuss gender, religion, culture and even socioeconomics in far more detail than we can do here.

Earlier in this book, we saw that some explanations of the crises are *wrong*, including the idea that the causes are always outside of capitalism, or the hypothesis that ecological destruction tends to create economic crises. We have also come across some explanations that are *weak* because they only dwell on the surface of crises: psychological explanations and those that focus on bubbles and speculation. Now we have arrived at explanations that are downright *dangerous.* We have come to racism.

The Eternal Answer

The capitalist class creates crises in its own image in a cocktail of nationalism and racism. Regardless of the context, regardless of what the crisis

---

59 Stuart Hall et al., *Policing the Crisis*, pp. 203–4, 320; Alexander Anievas and Richard Saull, 'Reassessing the Cold War and the Far-Right: Fascist Legacies and the Making of the Liberal International Order after 1945', *International Studies Review* 22, no. 3 (2020), pp. 370–95.

60 Poulantzas, *Political Power*, pp. 243, 288; Therborn, *What Does the Ruling*, pp. 190–5. See also Karl Marx, *The Eighteenth Brumaire of Louis Bonaparte*, in *Karl Marx: Surveys from Exile: Political Writings* (London: Penguin, 1973), pp. 143–249.

looks like or where and when it happens – somehow, racism always becomes a component. The crises are racialised from all sides: who causes the crises, who suffers, who is blamed and within what (post)-colonial patterns they are to be resolved. Crises are events within a capitalism that has been built for centuries on colonialism and imperialism. Racism is permanent enough to be present at every capitalist crisis, flexible enough to adapt to new crises, distinct enough to reproduce (old) power relations, and toxic enough to kill.

One group that is often blamed for everything that goes wrong, as we know, is the Jews. During the interwar period this had grotesque consequences, but Jews have also been blamed for everything from the Great Plague to the South Sea Bubble, the 1930s depression and the 2008 crisis. The anti-vaxxers during the Covid-19 pandemic were often attracted by racist conspiracy theories based on anti-Semitism. A limited study of twenty-seven anti-vaccine groups during the corona crisis found anti-Semitic conspiracy theories in 79 per cent of them.[61]

The idea that Jews rule the world through secret networks – the hallmark of anti-Semitism – works well with crises. When crises are events in which people are affected by seemingly uncontrollable phenomena with overwhelming consequences, it is certainly understandable that people search for *someone* who must have been in charge of this madness. My argument in this book – that crises are due to processes in capitalism and therefore not secretly planned – will never convince the conspiracy theorist. That it can never be proven that Jews have created any crises via their highly secret networks shows, apparently, just how incredibly adept they are at hiding their power.

Other groups have also been blamed throughout history. From a geopolitical perspective, it can be tempting to blame your enemies. The fact that Covid-19 began in China, for example, was of course a perfect fit for the American centre and right. In the economic crises of the 1970s it was difficult to blame the geopolitical antagonist, as the Soviet Union had a completely different economic system. Due to the oil component,

---

61 Richard J. Evans, 'Anti-Semitism Lurks behind Modern Conspiracy Theories', *Irish Times*, 16 February 2021; Magnusson, *Finanskrascher*, p. 90; Andreas Malm and the Zetkin Collective, *White Skin, Black Fuel: On the Danger of Fossil Fascism* (London: Verso, 2021), p. 517. See also Étienne Balibar, 'Racism and Crisis', in Étienne Balibar and Immanuel Wallerstein, *Race, Nation, Class: Ambiguous Identities* (London: Verso, 1991), pp. 218–20.

one could at least blame the Arab OPEC countries that were at war with Israel – 'the Middle East's only democracy'. Matt Huber shows how the 'oil crisis' in the United States was marked by anti-Arab xenophobia and how this found concrete expression in popular culture, satirical cartoons, opinion pieces and letters to politicians.[62] The racism also had a geopolitical undertone: how dare Muslim countries establish their own and independent energy policy! And where did the post-war golden age go? It was, of course, the Arabs who blew it all away.[63]

Racism is flexible and appears in different forms in different places. Under neoliberalism, there has been much implicit racism as countries in the Global South have had to conform strictly to Washington's models, while the conspiracies against George Soros show that anti-Semitism is anything but dead. When the Indonesian currency fell in 1997, sending Indonesia into the deepest of depressions, the blame was placed on the country's Chinese.[64] In 2008, the problems were because poor black people in the US had taken out too many loans, or because Greek workers were lazy. In Iceland, 'they' did not understand the risks of financial speculation and in Spain, 'they' should have slowed down the housing market in time. Italian finance minister Giulio Tremonti declared that the Italian banking system would get through the crisis just fine, because 'it did not speak English'. A few months later, the Italian economy was close to total collapse.[65]

We can say almost to the minute when the coronavirus created a 'global' crisis. It was not with the outbreak in China or the ravages in Iran. It was when the pandemic hit wealthy areas of northern Lombardy.[66] Since Donald Trump was the US president when the crisis broke out, we can safely say that the crisis would have had a racial explanation almost regardless of where it originated or what form it would take. That it now actually came from the main enemy, China, was just fuel for the fire.

---

62 Matt Huber, *Lifeblood: Oil, Freedom, and the Forces of Capital* (Minneapolis: University of Minnesota Press, 2013), pp. 103–8, 116.

63 See Stuart Hall, 'Gramsci and Us', in *The Hard Road to Renewal: Thatcherism and the Crisis of the Left* (1988), published on versobooks.com, 10 February 2017.

64 Klein, *The Shock Doctrine*, p. 264.

65 Tooze, *Crashed*, p. 73.

66 Andreas Malm, *Corona, Climate, Chronic Emergency: War Communism in the Twenty-First Century* (London: Verso, 2020), p. 21.

From Europe and much of the world came reports of the far right attacking Chinese, Muslims or others. As the regime in Hungary is less critical of China, there was no criticism coming from that direction, but Orbán seized the opportunity to be racist when it emerged that the country's first Covid-19 patient was an Iranian student. In Spain, the feminist 8 March demonstrators were first accused by the far right of spreading the virus. When it became clear that it was rather the anti-feminist 8 March event organised by the right-wing nationalist Vox that had spread the contagion on a large scale, Secretary General Ortega Smith tried to save face in classic national-chauvinist fashion by tweeting: 'Spanish antibodies will kick out the damned Chinese virus.'[67]

Racism may stomp out different paths, but stomp it does. *At the same time* as there were reports of racism against Asians in countries such as South Africa, Kenya and Nigeria, there were reports of increased racism against people of African background in Guangzhou, China.[68]

In Sweden, Johan Giesecke – a former state epidemiologist, consultant at the Public Health Agency, and part of the creation of the famous 'Swedish line' in 2020 – argued that Covid-19 was prevalent in Rinkeby, a poor and highly stigmatised city district in Stockholm, because of the culture there and because so many people did not know Swedish. At the same time, 'Swedes' hardly needed any rules at all, because 'Swedes' are so good at following the state's recommendations. (Both Giesecke, state epidemiologist Anders Tegnell and director-general of the National Public Health Agency Johan Carlsson broke with the state's own recommendations in 2020.) Ebba Busch, the leader of the Christian Democrats and since 2022 also Sweden's deputy prime minister, argued in 2020 that the spread of infection was particularly high among Somalis due to 'culture-specific reasons', that 'Somalis do not have the same tradition of written information or of medicine', and that 'people in vulnerable areas have lower trust in other people.'[69] This was not just a passive expression of everyday racism, but an active

67 Malm and the Zetkin Collective, *White Skin, Black Fuel*, p. 513. The Wikipedia page 'Xenophobia and Racism Related to the COVID-19 Pandemic' shows cases from over fifty countries.

68 Hsiao-Hung Pai, 'The Coronavirus Crisis Has Exposed China's Long History of Racism', *Guardian*, 25 April 2020; Jason Burke, Emmanuel Akinwotu and Lily Kuo, 'China Fails to Stop Racism against Africans over Covid-19', *Guardian*, 27 April 2020.

69 Ebba Busch, 'Våga tala klarspråk om corona och förorten', *Aftonbladet*, 2 April 2020, our translation.

production of racism during a crisis. Busch's comment came just when the crisis was in a state of shock and uncertainty – just when information was lacking, just when it was easiest to racialise the crisis and single out groups. Interestingly, when infection and mortality rates were lower in Rinkeby than in many other places during the two following waves, there were no culture-specific explanations.

In deep economic crises, voters often turn to the far right. Three German economists have studied how economic crises have affected politics in over 800 elections in twenty countries between 1870 and 2014. The study shows that people demonstrate and fight more after financial crises, but the main conclusion is that voters are attracted to parties that blame the problems on immigrants or minorities. Parties on the far right increased voter support by 30 per cent on average after financial crises. However, similar figures cannot be seen after what they call macroeconomic disasters or crises that have not also brought the banking sector into crisis. Our German economists use a different terminology than we do here, but, if we look more closely at the crises they call financial crises, they are – with the exception of the 1990s crisis – crises that have developed into deep organic crises: the 1870s–80s, the 1920s–30s and 2008.[70]

The issue of racism is one of the less explored dimensions of the climate crisis.[71] But like economic crises, the environmental issue has a long history of racism. Robert Miles goes back to Greco-Roman societies, where cultural differences were thought to be due to climatic, topographical and hydrographic conditions, and to the seventeenth and eighteenth centuries when the heat of the sun was thought to explain why some became darker-skinned, again creating inherent traits such as laziness. But if (negative) characteristics were due to geography, could people 'improve' by moving? This ambiguous racism disappeared with the introduction of scientific race theory in the nineteenth century.[72]

70 Manuel Funke, Moritz Schularick and Christoph Trebesch, 'Going to Extremes: Politics after Financial Crisis, 1870–2014', CESifo Working Paper, No. 5553, 2015, pp. 8–9, 33–4, Appendix D, pp. 53–5. See also Cervenka, *Vad gör*, pp. 78, 167.

71 Laura Pulido, 'Racism and the Anthropocene', in Gregg Mitman, Marco Armiero and Robert S. Emmett (eds), *Future Remains: A Cabinet of Curiosities for the Anthropocene* (Chicago: University of Chicago Press, 2020), p. 116; Malm and the Zetkin Collective, *White Skin, Black Fuel*, p. 321.

72 Miles, *Racism*, pp. 15, 29–30.

In *White Skin, Black Fuel*, we go back to the origins of fossil fuel combustion to describe how coal was considered the rightful natural resource of the 'white race', with which European colonisers could further subjugate both people and nature.[73] In human geography, environmental determinism became a strong trend beginning in the late nineteenth century, and it grew ever stronger until Europe was set on fire in the 1930s. Ideas developed that tied races to specific land areas and nations, or that associated a race's degree of civilisation with climate. With such 'science', German geographers could legitimise German expansion and their English colleagues legitimise the British Empire, while the North American geographer Ellsworth Huntington concluded that the optimal climate for building a civilisation was right where he happened to live – in Boston.[74]

Racism takes different forms in the climate crisis, including the pure racism of climate denial (is it all a fabrication from China?) and arguments that the Global North cannot accept immigrants because that would increase the global climate footprint.[75] We can more easily turn a blind eye to the crisis because it will hit non-white, poor and indigenous people first. One could imagine that it is impossible to blame poor and non-white people in poor countries for climate change. But the theory of overpopulation makes this possible too.

If it is not overpopulation, it is China. And China certainly has huge emissions and high growth. Then the picture is complicated by the fact that China's per capita emissions are still half those of the US, and it is said to be at the forefront of renewable energy sources. So, the arguments go back and forth, and climate debate becomes a contest of who is best and who is worst. Nation against nation – never class against class.

It is a fact that the climate crisis has been largely caused by white people in the Global North, and is disproportionately affecting

73 Malm and the Zetkin Collective, *White Skin, Black Fuel*, ch. 9.

74 Richard Peet, *Modern Geographic Thought* (Malden, MA: Blackwell, 1998), pp. 12-16; Arild Holt-Jensen, *Geography: History and Concepts* (London: Sage, 1999), pp. 42–5.

75 Pulido and colleagues discuss an interesting distinction between Donald Trump's explicit and spectacular racism and the silence that followed his ecological deregulations. They argue that the first – intentionally or not – concealed the second. See Laura Pulido et al., 'Environmental Deregulation, Spectacular Racism, and White Nationalism in the Trump Era', *Annals of the American Association of Geographers* 109, no. 2 (2019), pp. 520–32.

communities in the Global South, hitting them the hardest. The central class conflict is then between a (primarily white) capitalist class in the Global North and a (primarily non-white) working class – broadly defined – in the Global South. This very relationship is effectively obscured by geopolitics, nationalism and racism. At the climate summits, the leaders of the world's richest countries decide the fate of the world's poor. Only in a world where racism and class are fully normalised can such injustice pass as something perfectly routine. If we live in a Capitalocene, it is, as Françoise Vergès observes, a *racial Capitalocene.*[76] Classquakes also come as racequakes.

Racism hides class. Racism divides the working class by uniting sections of different classes in an imagined community. We should not reduce racism here simply to a process that divides the working class; Marxists have made that mistake before us. But racism is *also* a process that divides the working class. Racism is more than a class ideology coming from above, but it is also that.[77] When the ruling class has to discipline its workers – whether they are wage workers in factories, enslaved people on cotton plantations or 'employees' in the gig economy – racism is an effective tool. When people are to be forcibly relocated as a consequence of the new geographies of crises – either within cities or across larger geographical areas – racism emerges as a disciplining tool for crisis management. But racism is also alive within the working class.

In his analysis of racism in Britain, Stuart Hall saw it as an ideology that emerges from social processes – not least crises. For this ideology to become a real and historical political force, it had to be linked to 'the lived experiences of the "silent majorities"'.[78] Fears, tensions and anxieties created in/by crises can be distinctively and conveniently projected onto 'immigrants', 'race', religion or 'cultural differences'. Crises are often experienced and lived through racism.

When the (far) right is allowed to set the political agenda with

76 Françoise Vergès, 'Racial Capitalocene', in Gaye Theresa Johnson and Alex Lubin (eds), *Futures of Black Radicalism* (London: Verso, 2017), pp. 72–82.

77 Miles, *Racism*, pp. 89, 104; Stuart Hall, 'Racism and Reaction', in Commission for Racial Equality, *Five Views of Multi-racial Britain* (London: Commission for Racial Equality, 1978), pp. 23–35.

78 Stuart Hall, 'Racism and Reaction', p. 30; see also Stuart Hall et al., *Policing the Crisis*, pp. 333, 347, 394.

discourses on, for example, 'immigrants' or 'Muslims' *before* the crisis, this will be perceived as a natural go-to ideology when crises actually occur. Immigrants are scapegoats for recurring economic and social problems that accompany a system in crisis. (Other potential scapegoats mentioned by Hall and colleagues are communists, activists and 'foreign agitators'. Today we may add feminists and queer and trans activists.)[79] If you lose your job – as you might do in a crisis – your anger can be directed in many directions: at the boss who fired you, at the bourgeoisie at large or the rich, at the politicians, at your racialised neighbour or at 'immigrants' living in high-rises you have only seen on TV, or at feminists, or somewhere else . . . Which direction this takes depends largely on political struggle. After years of systematic stigmatisation of immigrants, obsession with the costs and problems of immigration, and a domestic nationalism in which the good and safe 'Swedish', 'Norwegian' or 'Danish' is perceived as the antithesis of the instability of the crises, it is not surprising that working-class anger is directed against other people in the same class, but of a different culture, skin colour, religion and so on.

When crises hit the racialised and disadvantaged working class hardest, they can sink into even deeper poverty, which, in turn, creates greater restlessness and alienation. This can lead to new expressions of anger, which can breed even more racism. And 'this is where the crisis bites', as Stuart Hall writes.[80] Immigrants and immigration become the scapegoats for all the problems of the crisis; racism becomes the solution.

Where crisis comes with chaos and uncertainty, racism can offer an imagined community of harmony, calm and stability. That this can be a good currency when the world is on fire should not surprise us. That it is, in fact, like pouring petrol on the fire is another matter.

Racism and nationalism not only tear apart social structures and the potential for working-class solidarity, they also legitimise reactionary political reforms that put their stamp on new hegemonic capitalist orders. While nationalism and racism are permanent – or at least latent – phenomena in every capitalist crisis, fascism comes only in very special situations. Where racism is the rule, fascism is the exception; if

79 Stuart Hall et al., *Policing the Crisis*, p. 50.

80 Stuart Hall, 'Racism and Reaction', p. 31.

racism is the eternal answer to crisis, fascism is the exceptional solution.

Which Crisis Leads to Fascism?

We can easily miss the role of fascism in the crises – and the role of the crises for fascism – if we stick only to the world of ideas. Perhaps the best-known definition of fascism is *palingenetic ultranationalism*, coined by the British historian Roger Griffin, the central feature of which is the core myth of national rebirth (palingenesis). This is an important concept, but it tells us very little about fascism as a historical force.[81] Many ideas central to fascism existed long before it was realised, such as ultra-nationalism, colonisation as a patriotic duty, anti-Semitism and the demonisation of Jews, and a preference for authoritarianism. If we, in contrast, are to understand fascism as a historical force, we cannot (only) ask the angry youth longing for a strong leader or the pensioner frightened by the 'immigrant' what they think about the world. We must (also) see within what historical conditions the potential of fascism can be realised.[82] The first thing we find then is that fascism was never for ordinary times.

As long as the political economy is relatively stable, fascism is dormant, and, according to many leading scholars of fascism, the triggering factor for a reversal is precisely *crisis*. If we truly want to understand the problem we must, according to Geoff Eley, 'begin by theorizing fascism in terms of the political crisis that produced it'. The social context, according to Robert Paxton, is 'a sense of overwhelming crisis beyond the reach of any traditional solutions'. Poulantzas argued that to understand fascism we must study 'both the political crisis to which the exceptional State is a response, and the particular kinds of political crises to which its specific forms correspond'. The masses would not have found fascism attractive, write

81 Roger Griffin, 'The Primacy of Culture: The Current Growth (or Manufacture) of Consensus within Fascist Studies', *Journal of Contemporary History* 37, no. 1 (2002), p. 24; Malm and the Zetkin Collective, *White Skin, Black Fuel*, ch. 7. On fascism as an ideology, a movement and a regime, see Ugo Palheta, 'Fascism, Fascisation, Antifascism', *Historical Materialism*, 7 January 2021, historicalmaterialism.org.

82 Nicos Poulantzas, *Fascism and Dictatorship: The Third International and the Problem of Fascism* (London: Verso, 2018), p. 39.

Malm and the Zetkin Collective, 'had they not felt the ground disappearing beneath their feet'.[83] If crisis is constitutive of fascism, the natural corollary question becomes: *Which* crises lead to fascism?

We must begin by noting that no crisis in itself has ever created fascism; it is merely an ingredient. When world history really shakes social relations to their foundations, it can also be unclear what is crisis and what is something else, such as social distress after a war. We need to see crises in context and context in crisis: Under what conditions can different types of crises contribute to fascism?

The interwar period brought crises of representation. This was a form of political crisis, which, according to Poulantzas, developed when big capital, the big landowners and later the petty bourgeoisie turned their backs on their political representatives. The crisis of representation contributed to the state of emergency and the rise of fascism, and, according to Poulantzas, went hand in hand with an ideological crisis.[84] Then there are economic crises. We see differences when comparing the economic development of the countries where fascism had the strongest hold. Germany was already a world-leading industrial nation when Italy began to industrialise. We cannot here escape the mass killing of the First World War, which traumatised European societies and produced veterans who formed the core of both *squadrismo* and the Freikorps.[85] Both countries experienced economic difficulties after the First World War and economic crises at the beginning of the 1920s; and, for German fascism, the yoke of war reparations was an important factor. Then came the Great Depression of 1929. Eley builds on Gramsci and Poulantzas when he sees the rise of fascism as a combination of crisis of representation and crisis of hegemony.[86] It is, of course, an analytical challenge to try to distinguish between different crises, since this becomes de facto an organic crisis,

83 Geoff Eley, 'What Produces Fascism: Preindustrial Traditions or a Crisis of a Capitalist State', *Politics and Society* 12, no. 1 (1983), p. 82; Poulantzas, *Fascism*, p.16; Malm and the Zetkin Collective, *White Skin, Black Fuel*, pp. 231–4. See also Richard Saull, 'Capitalism, Crisis and the Far-Right in the Neoliberal Era', *Journal of International Relations and Development* 18 (2015), p. 32; Gramsci, 'The International Situation', pp. 124–7.

84 Poulantzas, *Fascism*; on crisis and Germany, see pp. 100–8, and Italy, see pp. 123–31.

85 Saull, 'Capitalism, Crisis', pp. 32–4; Malm and the Zetkin Collective, *White Skin, Black Fuel*, pp. 231–2.

86 Eley, 'What Produces Fascism', pp. 78–9; see also Poulantzas, *Fascism*, pp. 25–35, 53–4, 71–2.

where crises not only coincide but also feed into each other, shaping and possibly reinforcing each other.

We cannot here undertake a thorough analysis of fascism, but we shall take three things into account in relation to crisis. First, fascism becomes a real possibility when the capitalist class does not seem able to create a crisis in its own image, and when states do not seem able to reproduce themselves as capitalist states. The crises discussed above came also within a context of a Russian revolution, and attempted revolutions in several of the core countries of capitalism. In a combination of political, ideological and economic crises, it seemed unclear whether the bourgeois states could really maintain social cohesion. Intellectuals from various traditions predicted the fall of capitalism. Fascism emerged when it seemed unclear whether the bourgeois states could guarantee the survival of capitalism at all.

Secondly, even if fascism was a popular movement with strong support among the petty bourgeoisie and also the working class, it would never have come to power if the ruling class had not accepted it. Previous attempts – Mussolini in 1919, Hitler in 1923 – had failed monumentally precisely because they were not sanctioned from the top of society. Fascism does not come to power through general strikes or social uprisings; it requires the support of the army and the police. The road to power is, on the one hand, through crisis and chaos, but, on the other hand, accompanied by an orderly transfer of decision-making power. In more stable countries like France, Belgium and Britain, the ruling class and political elite chose to reject the fascists' advances. Without the approval of the powers that be, fascism stands no chance, no matter how much ideology and propaganda or how many street militias fascists can mobilise. One classic analysis is that the finer bourgeoisie finds fascism vulgar and aesthetically repugnant, but enters into an alliance in the belief that the fascist movement can be tamed and controlled.[87]

Third, fascism is a way of reproducing capitalism through crises. Although fascist rhetoric and aesthetics may recall a perverted anti-capitalism – criticism of globalisation, finance and even of 'the

87 Malm and the Zetkin Collective, *White Skin, Black Fuel*, pp. 233–4; Poulantzas, *Fascism*, ch. 3; Rodrigo Campos, Sérgio Barcelos and Ricardo G. Severo, 'A Templar's Guide to Climate Change Denialism: The Brazilian Far-Right and Its Brand of "Cultural Marxism" Conspiracy', in Allen et al., *Political Ecologies*; Saull, 'Capitalism, Crisis', p. 35; Palheta, 'Fascism'.

universal' – it never contains a critique of the profit motive, private property or the fundamental class relations of capitalism. In other words, the critique of the universal, but never of what produces the universal, is a contradiction it can never move beyond. Similarly, fascism, in its rhetoric and ideology, can appear as a cross-class movement, because it emphasises the nation, corporatism and class collaboration, and has various forms of popular support. In reality it is dependent on the capitalist class, and stands firmly on the foundations of capitalism, thus affirming the dominance of the capitalist class over the working class.[88] At a deeper level, Ugo Palheta points out, fascism is an intensification of the exploitation and domination inherent in capitalism because it produces a social body that is extremely hierarchical (in terms of class and gender), normalised (in terms of sexualities and gender identities) and homogenised (in ethno-racial terms).[89] Imprisonment and massive crime such as genocide, Palheta points out, are not unintended consequences, but are always potentials of fascism. Fascism is an exceptional system of violence against those identified as enemies of the nation. And fascists are allowed into the corridors of power by (parts of) the ruling class in order to solve and manage political, ideological and economic crises – and thus ensure the continued existence of capitalism.[90]

What would fascism look like if it came to power in the Global North today? This is a difficult question. Discussions of fascism are often reduced to interwar Italy and Germany, but history never repeats itself and the Third Reich will never return. Rather than waiting for fascism as a ready-made package, we should look for trends pointing in its direction.[91] But we can also try to imagine different scenarios.

What would happen if a climate-denying, misogynistic and racist president lost an election, called it a conspiracy, rallied his armed street movement and mobilised his friends in the police, military and big business to overthrow the newly inaugurated president? Perhaps the

88 See, e.g., Saull, 'Capitalism, Crisis', pp. 30, 35, 38; Eley, 'What Produces Fascism', pp. 65–6; on the role of (large) farmers, see, e.g., Poulantzas, *Fascism*, ch. 6.

89 Palheta, 'Fascism'.

90 Malm and the Zetkin Collective, *White Skin, Black Fuel*, p. 235.

91 Or look for 'potentials'; see Eley, 'What Produces Fascism', p. 76; Malm and the Zetkin Collective, *White Skin, Black Fuel*, p. 236. On fascism as process, see Poulantzas, *Fascism*, pp. 65–7; Palheta, 'Fascism'.

president might have used his presidency to worship nationalism; to try to restore national pride and make his country *great* again (palingenetic ultranationalism); to celebrate the police's structural tendency to kill racialised people; to favour fossil capital; over time to fatten a dangerous street movement and to undermine the mass media. Perhaps this might happen in the midst of one of capitalism's greatest organic crises ever, including a huge economic crisis and a pandemic that provided opportunities to impose martial law. Perhaps this great nation might also be threatened by a geopolitical competitor as world ruler. What might happen then?

All the stars seemed aligned for Donald Trump on that night in January 2021 when his supporters stormed the Capitol. Everything pointed to fascism. But one key ingredient was missing. Dominant elements of the ruling class did not see the need to solve the crisis by extraordinary means.[92] There were other ways to solve the crises. For many capitalists, Trumpism was still too vulgar. Big-tech companies like Apple and Meta preferred Joe Biden, key parts of the state apparatus were not exactly allies (the FBI investigated him at times), while many – with good reason – thought Biden would be better at handling the corona pandemic.

One crisis that the twentieth century intellectuals did not discuss in relation to fascism was global warming. Is this another crisis with the potential for fascism? The climate crisis can develop in different directions, and, unfortunately, some of these do point towards fascism. We can identify three.

First, fascism could develop as a defence against progressive movements. Say popular pressure gets so strong that leading politicians actually start doing what it takes to keep warming below 1.5 degrees. That is, close oil rigs and coal mines; stop the airline industry; reorganise the geography of capitalism to minimise transportation; stop deforestation; and force the financial sector to invest in some sectors and ban others. Such a process would attack the freedom and power of key elements of the capitalist class and some sectors would be completely destroyed. Many would see this as a threat to capitalism itself. Is it

92 For discussion, see Richard Seymour, 'Is It Still Fascism if It's Incompetent?', *New Politics*, 8 January 2021.

inconceivable that large sections of the capitalist class would then turn to a fascist right that assured the world that the climate crisis was a conspiracy? Certainly not.[93]

Another path to fascism may be through the crises themselves – next-level shock doctrine. If crises mean opportunities for racism, then there will undoubtedly be more opportunities in the future. More and more heatwaves killing people and destroying crops, more and more wildfires, extreme weather pushing the sea miles inland, shortages of food and water, millions upon millions of people forced to flee and difficulties for the state apparatus to maintain the status quo. What if a socialist government proposed that the burdens should be shared equally? North American political scientist Cara Daggett writes about petro-masculinity: how unrest over gender structures and climate change can lead to a desire for increasingly authoritarian solutions. Daggett opens the question of whether the climate crisis will catalyse fascist desires for *Lebensraum*, a space for whites that is closed to the threatening others, whether they be immigrants or 'deviant' gender norms.[94] This is fascism as defence of imagined privilege.

A third input is variants of so-called eco-fascism. If the far right accepts that global warming is man-made, where will the responsibility lie? Who has destroyed the nature to which someone claims historical right? The far right will not blame fossil capital or corporations with white owners and headquarters in the Global North. During the Holocaust, 6 million Jews were killed based on absurd claims that they ruled the world. What if Jews (again), or Muslims (because it is the current fashion), or the Chinese (because of geopolitics), or Africans ('overpopulation'), or immigrants (or any other group you can mobilise hatred and racism against) are blamed for climate change? What happens if a fascist right decides to dispense justice in its own way? There is certainly cause for concern if we consider the experiences of the twentieth century in light of what climate scientists say the world will look like in the coming decades.

Crises are events that invite disinformation and conspiracies. It is frightening to think that more and more crises are coming at a time

93 See Malm and the Zetkin Collective, *White Skin, Black Fuel*, ch. 7.

94 Cara Daggett, 'Petro-masculinity: Fossil Fuels and Authoritarian Desire', *Millennium: Journal of International Studies* 47, no. 1 (2018), p. 44.

when the information society itself seems to be undergoing dramatic changes. Ernesto Araújo, Brazil's foreign minister under Bolsonaro, argued that globalisation is a conspiracy led by cultural Marxists operating through climateism, gender ideology and oikophobia.[95] The Araújo example is just one of many, but it points precisely to the political landscape we have to deal with in the near future.

Marxists often stress that fascism necessarily needs an alliance with big capital, and that fascism is a reaction against social movements and a counter-revolution against the growing socialist movement. These ideas have been criticised by non-Marxists, and rightly so when Marxists have reduced fascism to a mere question of class.[96] If we really want to understand the phenomenon in its full complexity, we must also understand misogyny, the relative autonomy of racism from class relations, ultra-nationalism and anti-liberalism, psychological aspects, and an uneven but combined development both regionally within countries and with the positioning of states within the imperialist world order, and more.[97] But we cannot understand fascism *without* emphasising its class character, how it relates to crises and the reproduction of capitalism.

A central hypothesis of this book is that capitalist crises reproduce capitalism. This does not necessarily mean that this always has to be the case, or that most people think this way. Fascism becomes the answer when crises call into question whether the ruling class and capitalist states can reproduce capitalism at all – at least within the framework of parliamentary representation and free speech. Fascism becomes a possibility when, in the face of crisis, the ruling class can only reproduce its world with extreme violence, martial law, extreme nationalism and extreme racism. *Fascism is a solution when it seems that the crises will* not *be able to reproduce capitalism.* In other words, fascism becomes a possibility when

95 For Brazil, see Campos, Barcelos and Severo, 'A Templar's Guide'; on cultural Marxism and climate deniers, see also Malm and the Zetkin Collective, *White Skin, Black Fuel*, pp. 300–13.

96 See, e.g., David Beetham (ed.), *Marxists in the Face of Fascism* (Chicago: Haymarket, 2019).

97 On international relations, see, e.g., Poulantzas, *Fascism*, pp. 22–4; Saull, 'Capitalism, Crisis', pp. 32–3. On fascism and relative autonomy, see Poulantzas, *Fascism*, pp. 83–8.

the basic hypothesis of this book is challenged. Fascism is the shock therapy when capitalism *really* needs to change in order to survive.

Fascism, then, is not a violation of the book's hypothesis. Rather, it is a last resort when the hypothesis is in danger of collapse.

Creative Destruction on Steroids

'What is fascism?' Antonio Gramsci asked himself in 1921, and replied, 'It is the attempt to resolve the problems of production and exchange with machine-guns and pistol-shots.'[98] The petty bourgeoisie, Gramsci continued, believes that gigantic crises can be solved with machine guns and military force. If the petty bourgeoisie really believes this, then we must agree with them. Economic crises *can* be solved with machine guns and war. Although militarisation is central to fascism, the relationship between war and crisis exists on many more levels. The combination of racism, nationalism and crisis creates constant tensions which, if allowed to develop, can lead to serious conflict. Matthew C. Klein and Michael Pettis point out in *Trade Wars Are Class Wars* how international trade conflicts can lead to armed conflicts, but observe that these are often born out of tensions and contradictions *within* countries, i.e., class struggles.[99] Along the same path, crises can develop into international conflicts and, in the worst case, war.

Wars are similar to capitalist crises in that they are often seen as random accidents arising from irrational mistakes, when in fact they are an integral part of capitalism. Both wars and crises are, of course, older than capitalism itself, but they have taken their modern forms by operating within our peculiar system of production. There is a huge literature on relations between war and capitalism, and we will not delve into it here, but we must have a brief discussion because this may – unfortunately – be very central to the development of crises in the years to come as well.[100]

Let us start with the economy. War undeniably poses several challenges to the economy: productive parts of the workforce are replaced

98 Antonio Gramsci, 'On Fascism, 1921', in David Beetham (ed.), *Marxists in the Face of Fascism* (Chicago: Haymarket, 2019 [1921]), pp. 82–3.

99 Klein and Pettis, *Trade Wars Are Class Wars*, ch. 2.

100 On economy and war, see, e.g., Mandel, *Late Capitalism*, ch. 9.

by unskilled labour; factories are forced to shift production; international trade declines; some companies go bankrupt; consumption may be rationed; lower unemployment may threaten profits. Furthermore, wars can destroy credit mechanisms and cause inflation, and states may have to increase their debt, to give a few examples. Wars can create a range of economic, social and humanitarian problems that may be related to crises, but need not be.

For those always seeking exogenous causes for crises, wars are often easy explanations. Wars do indeed affect the economy and might even amplify crisis tendencies, like the way the Vietnam War contributed to inflation in the 1970s. But the notion that wars *cause* economic crises does not hold as a general conclusion. Returning to our analysis in chapter 3, we can place the economic implications of war at abstraction level two – it affects how social formation is organised in time and space. Military conflicts may occasionally even *trigger* crises, but they do not *cause* them in any deeper sense. Rikard Štajner effectively refutes claims that the North American Civil War, the outbreak of the Russo-Turkish War in 1877 or the First World War *caused* economic crises. Instead, he suggests reversing the causal relationships: economic hardship and crises are more likely to provoke war.[101] In addition, wars are good for solving crises.

In the short term, war mobilisation and military production might turn economic hardship into a 'real or apparent boom'.[102] But wars can even contribute to solving crises at deeper levels. The Second World War remains the most famous example of a war that was partly caused by crises, and that also helped solve the crisis. In the 1930s, bankruptcies, stock market crashes, massive unemployment and state investments certainly contributed to creative destruction, but this was far from enough to solve the crisis. The economy was stuck in its problems – not unlike the situation in the early 2020s. The devastating Second World War was by far the 'best' round of creative destruction capital has ever witnessed. Baran and Sweezy therefore compared the two world wars – from an economic perspective – to 'epoch-making innovations'.[103]

101 Štajner, *Crisis*, pp. 189–200.

102 Ibid., p. 194.

103 Paul Baran and Paul Sweezy, *Monopoly Capital: An Essay on the American Economic and Social Order* (New York: Monthly Review Press, 1966), p. 220.

In addition to causing the direct destruction of capital – including landscapes that must be rebuilt – war often involves technological innovations. It can open up new markets (at least for the winners of the war) and, due to underinvestment in certain sectors such as housing for the civilian population, generate needs that can be met by the capitalist class or intervening states. War and militarisation can hinder critical thinking and some innovations, but foster other innovations which potentially can even help to bring the economy out of the crisis as a whole.[104]

We have previously discussed various reasons as to why many in the decades after the Second World War considered the era of crises to be over. Here is another: the arms race. Some argued that the problems of overproduction could be avoided by ever greater investment in military production, because here there was no problem of demand. Some argued that the tendency towards a higher organic composition of capital was counteracted when states invested in the military, because capital disappeared from sectors where this tendency was relevant. Still others argued that disproportionality between sectors producing means of production and consumer goods (departments one and two) was resolved when the military industry formed a third department that acted as a divider between the others.[105]

The arms race could be seen as a final solution to economic crises, but only if the military industry was perceived as a black hole outside the rest of the economy, where all problems could be absorbed.[106] Admittedly, this is a highly peculiar sector – producing death on assembly lines – but arms production and the military industry are also integrated into the economy and must be analysed as a relatively normal capitalist sector. Certainly, over-accumulated capital can find its way here, shaping the economy and the development of crises, but crises do not magically disappear. And, as for organic composition, the war industry, as we know, is characterised by high-tech innovations and, in general, a high organic composition. So, if anything, Mandel reminds us, increased

104 On how peace agreements can bring economic concerns, see, e.g., Štajner, *Crisis*, pp. 189–200, on Germany after the First World War; and Klein and Pettis, *Trade Wars Are Class Wars*, p. 56, for the Franco-Prussian War of 1870–71 and the 1873 crisis.

105 See Michael Kidron, 'Maginot Marxism: Mandel's Economics', *International Socialism* 36 (1969), p. 33; Baran and Sweezy, *Monopoly Capital*, ch. 7.

106 For discussion, see Mandel, *Late Capitalism*, ch. 9; pp. 278–302.

investment here leads to an acceleration of the trend in the economy as a whole.[107]

Ecological crises also have their links to war. They can be struggles over natural resources (like the guano war); metabolic rifts reinforced by new war technology (like the atomic bomb); wars over petroleum resources; wars creating a range of ecological problems with bombs, contaminated groundwater, toxins and much more. In addition to being a major environmental problem in their own right, modern wars also effectively overshadow the climate issue. Many also date the beginning of the Anthropocene to 1945, precisely because of atomic bombs.[108] Moreover, the military industry is probably the last sector that will stop using oil.

Most scientists seem to agree that climate change increases the danger of conflict and war, but then it gets more difficult. When scientists use different data, theories and methods, they can come to different conclusions.[109] The war in Syria is often discussed in this regard. A central argument is that climate change contributed to the extreme droughts of 2006/07 and 2008/09 that created water shortages and socio-economic problems, which in turn led to migration to the big cities, which then formed the basis of the 2011 popular uprising against the Assad dictatorship.[110] These are complex discussions. *How* important was climate change as a cause of war? If it was one of several causes, *how many* causes? If it is one of three, we can call it central; if it is one of a hundred, perhaps less central. And what is the relationship between climate change and the other factors? The main conclusion from the research seems clear: climate change may play a role in armed conflicts, but it is not the only factor, as countries do not automatically go to war when a crisis occurs. But there are compelling reasons to believe that the higher the average temperature, the more serious the situation may become.

107 Ibid., p. 285.

108 Sörlin, *Antropocen*, pp. 145–7.

109 For overview, see, e.g., Katharine J. Mach et al., 'Climate as a Risk Factor for Armed Conflict', *Nature* 571 (2019), pp. 193–7; W. Neil Adger et al., 'Human Security', in *Climate Change 2014: Impacts, Adaptation, and Vulnerability. Part A: Global and Sectoral Aspects. Contribution of Working Group II to the Fifth Assessment Report of the Intergovernmental Panel on Climate Change* (Cambridge: Cambridge University Press, 2014), pp. 755–91.

110 For different perspectives, see the forum on 'Climate Change and the Syrian Civil War' in *Political Geography* 60 (2017), pp. 232–55.

Climate change is therefore often referred to as a 'threat multiplier'.[111] How this develops depends on the political evolution of the crises.

Right now, with tensions rising between a (falling?) great power in the West and (a rising?) one in the East, with national chauvinism and racism festering in times of crisis, and with ecological crises further raising the level of conflict, we must – unfortunately – be especially aware that war may be crisis management by extraordinary means. Ecological crises will most likely contribute to more armed conflicts. The economic turbulence of the 2020s may continue as a long depression, and potentially exacerbate an organic crisis in a context of rising nationalism. Geopolitical balances of power remain unresolved. From a European perspective, relations with China and Russia are of paramount importance. Could a devastating war save capitalism again? Yes. Could the next world war serve as the creative destruction that capitalism needs to enter a new, hot, deadly and increasingly authoritarian phase of development? Absolutely.

The fact that crisis and war are two cataclysms of mankind, in Štajner's formulation, is a good entry point.[112] They have a similar class character: the broad working class has to pay an enormously high price, while at least parts of the capitalist class profit. The military can also be important in quelling any uprisings that might come out of capitalist crises. Both crises and wars discipline the working class, silencing opposition and marginalising ideas that stretch beyond the boundaries of the system. Short-term survival trumps radical thought. We saw earlier how (the threat of) unemployment and precarity become central components when a class creates a crisis in its image. Now we can add racism and national chauvinism, and, in extreme cases, even fascism and war. This does not mean that we should work from the pessimistic hypothesis that everything that can potentially get worse will do so. But it is imperative to carry this historical knowledge with us when we ourselves face the crises of capitalism.

111 Shehnoor Khurram, 'Boko Haram in the Capitalocene: Assemblages of Climate Change and Militant Islamism', in Allen et al., *Political Ecologies* , pp. 32–56.

112 Štajner, *Crisis*, p. 190.

6

Our Crisis Is Now

When Eugen Varga reflected on the world depression of the 1930s, he asked himself whether this could really be another normal crisis of capitalism: Was it just a transitory thing that would be followed by a boom, or was it, rather, an enduring crisis? Perhaps we had arrived at a more permanent crisis that would only be interrupted by short periods of boom?[1] The historical answer to Varga's question was clear: on the other side of a world war awaited the so-called golden age.

Almost a hundred years later, the question is again relevant. At the beginning of the 2020s, the crises seemed so numerous and interacted and converged in so many ways and spheres, that it, according to Zachary Levenson, was 'hard to figure out where to even begin to make an intervention.'[2] Corona crisis, climate crisis, financial crisis, debt crisis, economic crisis, ideological crisis, housing crisis, energy crisis, supply chain crisis, refugee crisis, new urban crisis, crisis of social reproduction, crisis of care, crisis of racism, political crisis, crisis of

1 Quoted in Rikard Štajner, *Crisis: Anatomy of Contemporary Crises and (a) Theory of Crises in the Neo-imperialist Stage of Capitalism* (Belgrade: KOMUNIST, 1976), p. 7. A plot twist came in 1946 when Varga argued that capitalism was perhaps somewhat more inherently stable than previously thought. This was highly unpopular within Stalinism, so Varga was dismissed as director of the Moscow Institute of World Economics and World Politics, and in the second edition of the *Great Soviet Encyclopedia* he was reduced to a 'bourgeois economist'.

2 Zachary Levenson, 'An Organic Crisis Is upon Us: When Gramsci Goes Viral', *Spectre*, 20 April 2020.

democracy, legitimation crisis, livelihood crisis, personal security crisis and more . . . After Russia attacked Ukraine in 2022 there was a geopolitical crisis, an energy crisis, an inflation crisis, a cost-of-living crisis, and, additionally, the war itself was often called a crisis.

Discussions around 2020 were often about dual crisis or triple crisis, or 'several nested crises', or crisis chains or inter-crisis relationships.[3] Crisis was hot. The Collins Dictionary word of the year in 2022 was 'permacrisis', described as 'an extended period of instability and insecurity'.[4] A similar and much more popular buzzword also appeared in 2022: 'polycrisis'. It was Jonathan Derbyshire's 'year in a word 2022' in the *Financial Times*; it reverberated at COP27 in Sharm el-Sheikh 2022 and was, according to Susan Geist, copywriter for McGregor Boyall, 'by far the most used phrase on everyone's lips' at the World Economic Forum 2023.[5] Polycrisis had previously been articulated by former president of the European Commission Jean-Claude Juncker, and was celebrated in 2022 by former US Treasury secretary Larry Summers. If anything, the popularity of the concept indicates that even parts of the ruling class and leading policy makers – not least in the Global North – were anxious about the future.

According to Adam Tooze, a main advocate of the concept, polycrisis is a situation that involves multiple crises and where 'the whole is even more dangerous than the sum of the parts'.[6] In this work, we have identified some weaknesses in the crisis literature; for example, lack of

3 Triple crisis, see Joseph Choonara, 'A Triple Crisis', *International Socialism* 167 (2020), pp. 1–51; Michael Jacobs, 'Capitalism Is in Crisis. And We Cannot Get Out of It by Carrying On as Before', *Guardian*, 8 November 2019. For nested crises, see Salar Mohandesi, 'Crisis of a New Type', *Viewpoint Magazine*, 13 May 2020. For crisis chains/inter-crisis relationships, Annika Bergman-Rosamond et al., 'The Case for Interdisciplinary Crisis Studies', *Global Discourse: An Interdisciplinary Journal of Current Affairs* 12, no. 3–4 (2022), pp. 465–86.

4 David Shariatmadari, 'A Year of "Permacrisis"', *Collins Dictionary*, 1 November 2022.

5 Susan Geist, 'What Is a Polycrisis, Why Is Everyone Talking about It and How Could It Affect Your Business?', mcgregor-boyall.com, 22 March 2023; Jonathan Derbyshire, 'Year in a Word: Polycrisis', *Financial Times*, 1 January 2023.

6 Adam Tooze, 'Defining Polycrisis: From Crisis Pictures to the Crisis Matrix', adamtooze.com, 24 June 2022. Tooze argues a problem becomes a crisis when it challenges our ability to cope and thus threatens our identity, and at times one might feel as if one is losing one's sense of reality. See also Ville Lähde, 'The Polycrisis', aeon.co, 17 August 2023.

definitions and vagueness about how crises are related. Polycrisis reproduces both these problems. Concerning definition and demarcation, we can just note that quite different problems, stressful events, emergencies and destabilising factors are included as components in polycrisis. Tooze includes, for example, the hunger crisis, nuclear escalation and Trump coming back to power, while the World Economic Forum includes widespread cybercrime and cyber insecurity, terrorist attacks and geoeconomic confrontation.[7] Concerning relations between the crises, they are generally dealt with at the level of appearances; focus is on how the effects from one crisis influence another. This is crises as events, or what we defined in chapter 3 as abstraction level one, or exceptionally at level two. No wonder the concept has been criticised, not only for merely being a catchy way of saying that several things are happening simultaneously, but also for refusing to acknowledge capitalism as something conditioning these various 'crises'. Daniel Drezner argues polycrisis mostly seems like a 'device to make people care about the Really Bad Things that climate change can do, without turning people off by warning them yet again about the hazards of climate change'.[8]

Ville Lähde discerns two uses of polycrisis: the first pertaining to 'a' polycrisis, which points to a general situation where different crises interact and strengthen each other, and the second concerning 'the' polycrisis, which aims to characterise our specific moment in world history.[9] Concerning the former, we should always try to articulate as precisely as possible which crises we are actually talking about. When we need general concepts to describe situations where crises interact and strengthen each other, I think the Gramscian concepts of organic and hegemonic crises already do the work. Concerning the latter, on naming the actual situation of interacting crises in the 2020s, I hope – if this needs a label of its own – that someone can come up with something better than polycrisis.

---

7 Adam Tooze and the World Economic Forum use both 'crisis' and 'risk': I read this as risks being something that could become crises, or potential problems that could turn very serious. See Tooze, 'Defining Polycrisis'; World Economic Forum, *The Global Risks Report 2023: 18th Edition Insight Report* (Geneva: World Economic Forum, 2023).

8 Güney Işıkara, 'Beating around the Bush: Polycrisis, Overlapping Emergencies, and Capitalism', *Developing Economics*, 22 November 2022; Daniel Drezner, 'The Concept of "Polycrisis" Was Everywhere in Davos. But Is It Saying Anything Meaningful?', *Vox*, 28 January 2023.

9 Lähde, 'The Polycrisis'.

Permacrisis and polycrisis might be problematic concepts, but what they seek to describe is real. There *is* much turbulence. Several crises interact and create turmoil, and they interact with turbulent processes that fall outside our definition of crisis, and these converge, interact and create a common and more serious situation than the individual parts could have done alone. As the turbulence continues, it is tempting to ask Varga's question again: Is crisis still the exception, or has it become the new rule? With dark clouds on the horizon, it is tempting to muddle things up and conclude that everything will go to hell and nothing will be the same. If we are to learn any lessons from the history of capitalism, it is that we must take organic crises deadly seriously, but also maintain nuances by distinguishing between the different crises.

When breaking things down, we see that not everything is new. The crises that crippled the world in the early 2020s are all ordinary in the sense that the ingredients are well known. Surely, the lockdowns and the diabolic character of the 2020 crisis were fairly novel, but we have seen most of the components before. Learning from history, it is plausible to say that the economic crisis will pass when the creative destruction has done its work. New hegemonies will replace old. There will be new pandemics, but Covid-19 will pass. The Russia–Ukraine war will end, with brutal human consequences and massive profits for arms dealers and oil industries. And new business opportunities for many. What makes Varga's question more complex today than in the 1930s is the climate crisis. This is new territory. When crises and wars return, they will arrive in a warmer world. Here, we need more good analysis and powerful political interventions, not primarily new concepts.

It is impossible to grasp the organic and hegemonic crises in the 2020s without understanding the 2010s. The decade when nothing happened, but everything was in motion. When problems were postponed, but never solved. Particularly galling from a left-wing perspective was that there was no shortage of reasons for optimism. But, every time there seemed to be an opportunity for success, everything went wrong. Movements that could promote social and political justice were dashed: Occupy in the US, Syntagma Square and Syriza in Greece, M15 and Podemos in Spain, the Arab Spring, Corbyn and Sanders. If the 2010s were worthless from a political and economic perspective, this cannot be compared to how disastrous the decade was concerning global warming.

The Useless Decade (2008–20)

The crises of the 2020s came after a decade when the economic crisis could have been resolved and we could have started to reduce emissions. But politicians and others in power did everything they could *not to* reorganise the political economy. They refused to leave a framework that was already out of date. Politicians were unwilling or unable to acknowledge that the 2008 crisis was a hegemonic crisis of capitalism, and so it could not be solved.

The statement that the crisis could not be resolved needs to be clarified, as such a formulation depends on definition and delimitation. If we define the economic crises narrowly, we see a crisis in 1929–33, another in 1937–39, a crisis in 1973–74 and another in 1979–81. If we see the crises as hegemonic crises, it is clear that the problems were not solved in 1933 and 1974, which allows us to talk about the crises of the 1930s and 1970s respectively. The 2010s tie together crises before (2008) and after (early 2020s) the decade, which together constitute a hegemonic crisis. A hegemonic crisis is, for Stuart Hall, an accumulation of contradictions. It marks a profound rupture in the political and economic life of society, and is the moment when political leadership and cultural authority are exposed and challenged. Poulantzas sees hegemonic crises as situations in which neither classes nor class fractions are able – by themselves, through various organisations or through the state – to impose their new power blocs on the other classes and class fractions.[10]

If we want to understand *how* this happened, and why underlying problems were never solved but postponed to the future, we must beware of making a mess of the whole and then getting lost in the swamp of ambiguity. We must be cautious to ensure that concepts like organic crisis, hegemonic crisis and polycrisis do not contribute to further obscurity. This is often called the crisis of neoliberalism, but 'neoliberalism' is too broad a concept to start our analysis from. Let us start by breaking down and discussing the 2010s from four different vantage points: politics, economics, ideology and environment.

10 Stuart Hall et al., *Policing the Crisis: Mugging, the State and Law and Order* (London: Macmillan, 1978), ch. 7, esp. p. 217; Nicos Poulantzas, *Fascism and Dictatorship: The Third International and the Problem of Fascism* (London: Verso, 2018), p. 72.

Lifting the Burden from the Shoulders of the Rich and Placing It on the Shoulders of the Working Class (Political Argument)

An emblematic moment in 2008 was the failure of Lehman Brothers, but equally emblematic was the fact that many banks were *not* allowed to fail. When banks found it difficult to borrow from other banks, it was deemed necessary to bail out the banks to save the system.

Politicians in the EU were initially more defensive than in the US, but when both the euro and the EU as a political project appeared to be at stake, Central Bank chief Mario Draghi radically reversed course. The European Central Bank proclaimed in 2012 that it was willing to do *whatever it takes* to defend the euro. This reassured the market, at least to some extent.[11] States set up stability plans with associated stability funds, the idea being that the state and banks would work together to save banks from future failures. The message from the states was clear. If finance capital were to blow up again and trigger a crisis, the state would once again bail out the banks. After 2008, many states in the Global North launched various investment programmes. Globally, stimulus packages of around 2 per cent of global GDP were set up.[12] Suddenly there was talk of a return of Keynes, and Joseph Stiglitz euphorically claimed that 'we are all Keynesians now'.[13] To what degree this was Keynes returning remains an open question, but at least it is safe to say that state spending was back – and on a scale hardly seen before.

Concerns about government deficits were irrelevant when governments needed to rescue the financial system. Once the system was saved, however, old dogmas were reverted to. This annoyed Keynesians like Paul Krugman, who argued that the reason Keynesian policies did not solve the crisis was that investment was too low, and the money primarily went to bailing out the banks rather than stimulating other parts of

11 Mario Draghi, 'ECB's Draghi to the Euro's Rescue?', *Euronews*, published on YouTube, 26 July 2012; Adam Tooze, *Crashed: How a Decade of Financial Crises Changed the World* (New York: Viking, 2018), ch. 18.

12 For global figures, see Frank Hoffer, 'The Great Recession: A Turning Point for Labour?', *International Journal of Labour Research* 2, no. 1 (2010), p. 100.

13 Joseph Stiglitz, 'Getting Bang for Your Buck', *Guardian*, 5 December 2008; 'John Maynard Keynes. The Keynes Comeback', *Economist*, 1 October 2009, unsigned; Robert Skidelsky, *Keynes: The Return of a Master* (New York: Public Affairs, 2009); James Evans, Phil Jones and Rob Krueger, 'Organic Regeneration and Sustainability or Can the Credit Crunch Save Our Cities?', *Local Environment* 14, no. 7 (2009), p. 683.

the economy.[14] The fact that OECD countries had increased their gross public debt from 92.5 per cent of GDP in 2009 to 103 per cent in 2011 became an argument for a resolute return to austerity policies.[15] If only governments could get their debts under control, as the saying went, companies would be able to invest again, consumers would consume again and the economy would recover. That did not happen.

While states conducted austerity measures with one hand, they continued to inject money into the system to stimulate economic growth with the other. Two main strategies here were quantitative easing (central banks purchasing securities to increase the money supply) and lowering the repo rate (encouraging the creation of more debt). Many central banks tried to revive the economy through what would be called 'the greatest monetary experiment in history'.[16] The underlying assumption was that the system itself was healthy; it just needed more money to get going. However, the driving force of capitalism is not money supply, innovation or any creative power – it is profit. Money sought out sectors that yielded profits, which was often speculation on stock markets and in real estate.

Additionally, the Chinese response to the 2008 crisis was of world-historic proportions. No country has ever implemented a more comprehensive counter-cyclical crisis programme. Similar mobilisations in the Global North have only occurred when countries have been threatened by war. The gigantic stimulus spending is said to have amounted to over 19 per cent of GDP. There was housing construction (or rather city

14 Paul Krugman, *End This Depression Now!* (London: W. W. Norton, 2012); 'The Austerity Agenda', *New York Times*, 1 June 2012.

15 Figures from OECD, *Restoring Public Finances, 2012 Update* (Paris: OECD Publishing, 2012), p. 13; for discussion see Hoffer, 'The Great Recession', p. 103. On austerity before 2008, see, e.g., Terry Nichols Clark, 'Old and New Paradigms for Urban Research Globalization and the Fiscal Austerity and Urban Innovation Project', *Urban Affairs Review* 36, no. 1 (2000), pp. 3–45; Naomi Klein, *The Shock Doctrine: The Rise of Disaster Capitalism* (London: Penguin, 2007). On austerity after 2008, see, e.g., Derek Wall, *Economics after Capitalism: A Guide to the Ruins and Road to the Future* (London: Pluto, 2015), ch. 2; Michael Roberts, *The Long Depression: How It Happened, Why It Happened, and What Happens Next* (Chicago: Haymarket, 2016); Betsy Donald et al., 'Austerity in the City: Economic Crisis and Urban Service Decline?', *Cambridge Journal of Regions, Economy and Society* 7 (2014), pp. 3–15; Jamie Peck, 'Austerity Urbanism', *City* 16, no. 6 (2012), pp. 626–55.

16 Andreas Cervenka, *Vad gör en bank?* (Stockholm: Natur & Kultur, 2017), p. 12, our translation.

construction) of extreme proportions, huge investments in high-speed rail and the largest single investment in health care the world has ever seen. Between 2011 and 2013, China used 40 per cent more cement than the US did in the entire twentieth century.[17] We will not take the geopolitical analysis any further here, but the importance of China's efforts in the 2010s to make the leap into a highly modern and developed economy, and shift from export-led to more domestically generated growth, cannot be emphasised enough.[18]

Political crises come in different forms. We saw that Poulantzas discussed the crisis of representation concerning the 1920s/30s when confidence in representative democracy disappeared. We can talk about 'Pasokification' when individual parties break down within a system, and, with Gramsci, we can talk about authority crisis or legitimation crisis when hegemonic views start to crumble and when the ruling class's definition of the common good is severely challenged.[19] One form of political crisis that was central in the 2010s was the crisis of crisis management.

Just as the Great Depression could not be ended with more laissez-faire or stagflation could not be resolved with Keynesianism in the 1970s, the crisis of neoliberalism could not be solved with cheaper money, further privatisation, austerity for the working class, tax cuts or higher real estate prices. But, if this crisis management was *not* pursued, a very deep crisis would definitely be a fact. There were no more working tools in the neoliberal toolbox.

---

17 David Harvey, 'Realization Crises and the Transformation of Daily Life', presentation at the Korea Press Center in Seoul, South Korea, 21 June 2016, available at magazine.changbi.com; Tooze, *Crashed*, pp. 7, 86–7, ch. 10; Perry Anderson, 'Situationism à L'envers?', *New Left Review* 119 (2019), p. 71. For example, between 2008 and 2014, the rail network capable of travelling above 250 km/h was expanded from 1,000 to 11,000 kilometres.

18 On how US hegemony is reinforced by crises, since the dollar is still the currency everyone wants in crises, see Ann Pettifor, *The Case for The Green New Deal* (London: Verso, 2020), pp. 24–6; Tooze, *Crashed*; Adam Tooze, *Shutdown: How Covid Shook the World's Economy* (London: Allen Lane, 2021), pp. 116–17.

19 Antonio Gramsci, *Selections from the Prison Notebooks* (London: Lawrence and Wishart, 1971), pp. 206–76; see also Stuart Hall, 'Gramsci and Us', in *The Hard Road to Renewal: Thatcherism and the Crisis of the Left* (1988), published on versobooks.com, 10 February 2017; Levenson, 'An Organic Crisis'.

The Expected Surprise (Economic Arguments)

The 2020 crisis is radically different from all major economic crises in capitalism in that 'everyone' saw it coming. Months before the crisis was triggered by the pandemic, the question was: What will the crisis be like this time? But no one could foresee the actual form it would take.

Many people find it hard to see crises coming, probably because they (unconsciously) do not want to see them, because they are in places in society where capitalism is often a bed of roses, or because they themselves may have been involved in regulating the economy and have pride in their construction. Since the norm is slow and steady growth, forecasters tend to expect the same next year. According to a study by the *Economist*, which compiled 100,000 economic forecasts of GDP across fifteen rich countries, economic forecasts may work relatively well for short periods (i.e., a few months) but quite poorly over years.[20] Crises are obviously difficult to predict. In any case, experts and politicians tend to imagine that right now, right here, growth and profits will just roll on. Even when the crisis has started, forecasts tend to be overly optimistic.

The case in point is when US president Calvin Coolidge assured the US Congress in December 1928 that no Congress had ever 'met with a more pleasing prospect than that which appears at the present time'. Just before the stock market crash of the following year, the country's experts assured everyone that the US had entered a new era of 'enduring prosperity'.[21] In the summer of 1929, the Harvard Economic Society declared that business was going well; in January 1930, they claimed that the most serious stage of the crisis had passed; in June, they proclaimed that the turnaround was in sight, and in November, that the downward trend was over. A year later, the Harvard professors stopped issuing forecasts.[22]

Not many foresaw the crisis of the 1970s. Although the rate of growth of world exports slowed towards the end of the 1960s and inflation and unemployment began to rise, most people were still optimistic.[23] According to

20 'GDP Predictions Are Reliable Only in the Short Term', *Economist*, 15 December 2018.

21 Speech available at infoplease.com.

22 John Kenneth Galbraith, *The Great Crash 1929* (London: Penguin, 2009), ch. 2; see also Štajner, *Crisis*, p. 67.

23 On how economists were too optimistic in the early 1970s, see Štajner, *Crisis*, ch. 3.

Alan Greenspan – chairman of the US Federal Reserve from 1987 to 2006 – no forecaster of note could have predicted the financial crisis of 2007–08. In September 2007, just days before the bank Northern Rock needed to ask the government for emergency loans, David Cameron, three years before he himself would become UK prime minister, said the world economy was more stable than it had been for a generation. Five months later, Northern Rock was nationalised.[24] Credit rating agencies gave top ratings to completely rotten mortgages because they had based their forecasts on the price performance of such loans over the previous twenty years. Those who saw dark clouds in the sky before 2008 were often looking in the wrong direction. Many saw US sovereign debt to China and dependence on Chinese credit as the main source of instability.[25]

Optimism continued in the beginning of the 2010s, with Marxists often being the exception. When Michael Roberts claimed in 2009 that the economy would bottom out in 2018, this seemed to many like a classic Marxist doomsday prophecy. Today, we can argue that even Roberts was not pessimistic enough.[26] The optimistic forecasts repeatedly clashed with reality. By 2015, the IMF was already questioning how smart the austerity policy had actually been. The UK's former business secretary Vincent Cable warned in 2016 that the world was building up risks similar to those before 2008.[27]

In 2018 and 2019, it became increasingly mainstream to be bearish. Mervyn King, England's former central bank governor, said in 2019 that by sticking to monetary policy and pretending the banking system was safe, we were sleepwalking towards another crisis. In April that year,

---

24 On Cameron, see Michael Roberts, 'The Failure of Forecasting', *The Next Recession*, 24 December 2013. On Greenspan, see Alexandra Wolfe, 'Greenspan: What Went Wrong', *Wall Street Journal*, 18 October 2013. On Northern Rock, see BBC, 'Northern Rock to Be Nationalised', BBC, 17 February 2008, unsigned. On how some economists were concerned about the risk of crisis, see Carlota Perez, 'The Double Bubble at the Turn of the Century: Technological Roots and Structural Implications', *Cambridge Journal of Economics* 33, no. 4 (2009), pp. 779–805.

25 See Tooze, *Crashed*, p. 41.

26 Roberts, *The Great Recession*; Roberts, 'The Failure of Forecasting'; Michael Roberts, 'Forecast 2020', *The Next Recession*, 20 December 2019.

27 On IMF 2015, see Stephen Grenville, 'IMF Belatedly Rethinks Austerity', lowyinterpreter.org, 29 June 2015. For Cable, see Ewa Persson, 'Brittisk politiker: Ökade risker för ny finanskris', *Arbetsvärlden*, 27 January 2016. Ben Wright argued in the *Telegraph* (23 March 2015) that 'the world's next credit crunch could make 2008 look like a hiccup'.

Harvard Business Review wrote that businesses should prepare for the next downturn. *Business Insider* wrote in August 2019 that three of Europe's largest economies – Germany, Italy and the UK – were already in recession, pointing out that with repo rates at negative, central banks had no more tools to solve the problems. In the same month, Reuters reported that both General Motors and Ford had begun planning for a possible crisis. In September 2019, *Forbes* published the article 'The Coming Recession: Be Smart When Others Are Fearful'. In a similar tone, the Confederation of Swedish Enterprise was able to state in 2019 that 'there is currently great uncertainty about how long and how deep the global economy will fall'.[28] The US Federal Reserve worried about the rise in corporate debt, even outside the financial sector, and in the OECD debt had risen to an 'all-time high'. The period of falling interest rates coincided with a steady increase in debt in the world's major economies. There were warnings of increased risk-taking in the corporate sector, and government debt increased in the 2010s in most countries.[29] It became increasingly clear that large corporations and banks had soaked up cheap credit during the 2010s, primarily to buy up stocks and bonds, which resulted in sharply rising stock prices. The productive economy stood still.

The 2010s were not a good time for optimists, but neither were they a time for prophets of doom. In fact, according to Michael Roberts, the decade was the first since 1945 without major downturns in the

28 Larry Elliott, 'World Economy Is Sleepwalking into a New Financial Crisis, Warns Mervyn King', *Guardian*, 20 October 2019; Martin Reeves, Kevin Whitaker and Christian Ketels, 'Companies Need to Prepare for the Next Economic Downturn', *Harvard Business Review*, 2 April 2019; Yusuf Khan, 'Three of Europe's Biggest Economies Are Probably in Recession – and the ECB Is Out of Bullets', *Business Insider*, 10 August 2019; Ben Klayman, 'Ford Planning for Possible Economic Downturn: Executives', Reuters, 13 August 2019; Grant Freeland, 'The Coming Recession: Be Smart When Others Are Fearful', *Forbes*, 9 September 2019. For Sweden, see Torbjörn Halldin et al., *Det ekonomiska läget. November 2019*, Confederation of Swedish Enterprise (Svenskt Näringsliv), p. 6. For an overview of critical scholars who saw the crisis coming, see Eric Toussaint, 'The Capitalist Pandemic, Coronavirus and the Economic Crisis Part 1', cadtm.org, 19 March 2020.

29 S. Çelik, G. Demirtaş and M. Isaksson, *Corporate Bond Market Trends, Emerging Risks and Monetary Policy* (Paris: OECD Capital Market Series, 2020), pp. 3, 5. For Sweden: Halldin et al., *Det ekonomiska läget*, p. 8; Göran Therborn, *Kapitalet, överheten och alla vi andra* (Lund: Arkiv förlag, 2018), pp. 50–9. On Swedish housing debt, see Timothy Blackwell, 'Manufacturing Debt: The Co-evolution of Housing and Finance Systems in Sweden', PhD diss., University of Sussex, 2019; Cervenka, *Vad gör*, pp. 42, 47–50, 71.

dominant economies. At the same time, it was also the weakest post-recession recovery since 1945. Falling unemployment in many countries helped keep the economy afloat. Rather than investing in new technology and productivity gains, many companies profited from increasingly cheap labour, the foremost example being highly pressured gig workers.[30] Rather than rounds of creative destruction, companies profited either on the stock market or by hiring workers with zero-hours contracts and no benefits. Or a combination: some invested in companies whose business idea it was to be brutal to workers, like Uber or foodora. Many could get a job, but often without a living wage.

The 2010s were characterised by the remarkable combination of (relatively) low unemployment and low interest rates, but without high inflation. The economy simply would not fly. Where forecasts were previously overly optimistic, in 2019 they were not pessimistic enough. In March 2019, the OECD lowered its already low expectations from 1.8 per cent to 1 per cent growth for 2019 and from 1.6 per cent to 1.2 per cent for 2020.[31] Then came 2020, when everything became highly extraordinary, and even a modest 1 per cent growth became a distant dream.

If the pandemic had not triggered the 2020 crisis, then something else (probably) would have. For example, as the coronavirus began to spread, the West saw a dramatic fall in the price of oil when Russia refused to cut production. This could have triggered the crisis. Then everyone would have been talking about a new oil crisis, and the powers that be would have said that once the geopolitical problems were resolved, the economy would recover. Some pointed out that a war between the US and Iran could trigger the crisis, while others emphasised the trade conflict between the US and China.[32] It is also possible to imagine that Brexit, a total breakdown of the Italian economy or the high levels of private and public debt would trigger a major international crisis; then everyone would have been talking about the trade crisis, the Brexit crisis, the Italian crisis or the debt crisis.

30 Roberts, 'Forecast 2020'.

31 Delphine Strauss, 'OECD Slashes Eurozone Economic Growth Forecasts', *Financial Times*, 6 March 2019.

32 On Iran, see Toussaint, 'The Capitalist Pandemic'. On China, see Klayman, 'Ford Planning'. On naming crises, see also Richard D. Wolff, *The Sickness Is the System: When Capitalism Fails to Save Us from Pandemics or Itself* (New York: Democracy at Work, 2020), p. 22.

When We Stopped Believing (Ideological Arguments)

During the useless decade, many of the crisis's morbid symptoms were visible in the media. On the news, broadcasters could enthusiastically talk about rising house prices and the banks' fantastic surpluses. Then it was back to normal news and old problems. I will not make a detailed analysis of the ideological crisis, but I will briefly point out some movements in the world of ideologies because they have major implications for our organic crisis. In this very limited discussion, we will only see ideology as a way for rulers to legitimate their power in the world of discourses, theories and ideas.[33]

A central neoliberal ideological promise was that privatisation, deregulation, lower taxes and free trade – all in the beautiful name of freedom – would benefit *everyone*. If this was not the case, then all that was needed was one more reform, one more tax cut or one more privatisation, and *then* the hypothesis would be realised. In *No Is Not Enough*, Naomi Klein has a most interesting passage about how Donald Trump's appearances in *The Apprentice* testified to the ideological and intellectual crisis of neoliberalism.[34] Whereas, in the 1990s, the neoliberal message was that everyone benefitted from globalisation and the free market, Trump's message was that the world is a rotten place crowded with poor people and losers, but also a place where *you* can be a winner. This symbolises very well the difference between optimistic neoliberalism and the dark conservative winds that have been replacing it.

A dying ideology quickly feels ancient. Today, it is mainly dogmatists who still believe in what was the hegemonic perspective at the turn of the millennium. Who today believes that tax cuts for the rich will benefit the poor? Who believes that cities, by attracting the rich, also gain an 'entrepreneurial spirit' that will trickle down? Or that we best provide housing for the poorest by building for the richest?[35] Who believes today that health care will be better in poor and stigmatised neighbourhoods if the local health centre is privatised? Who believes that New Public

33 On ideological crisis, see also Poulantzas, *Fascism*, pp. 76–8.

34 Naomi Klein, *No Is Not Enough* (New York: Haymarket Books, 2017), ch. 3.

35 On Richard Florida, who for almost twenty years preached about the creative class and the creative city, admitting his theories more than anything created segregation and social polarisation, see Richard Florida, *The New Urban Crisis* (New York: Basic Books, 2017).

Management will make the public sector more efficient? That more power to banks and financial institutions will stabilise the economy, or that higher house prices and higher debt will drive growth in the future? Does anyone still believe that poor countries will be lifted out of poverty by global free trade? Only fools could think that the 2020 pandemic could best be handled by more privatisation, more free trade, austerity and just-in-time production.[36] In the spring of 2021, even US president Joe Biden admitted that 'trickle down' had never really worked.[37]

The ideological crisis is a social paroxysm. Ideological crises are never purely discursive phenomena, but always relate to the object in question. On the surface, we stop believing stories about privatisation creating wealth. At deeper levels, it is about faith in a broader neoliberal ideology that is losing its legitimacy because what it is supposed to legitimise – the economy and politics – is in crisis. Ideological crises must be seen as the relationship between what the ideology promises (growth, progress, freedom) and its underlying structures (low growth, bubbles, segregation).

Neoliberal ideas entered a deep crisis during the 2010s, but the class power that the ideas were supposed to legitimise remained. Social problems such as poverty, segregation and social polarisation – which all remained or increased – needed to be legitimised in new ways. When poverty could no longer be explained away by the fact that it would soon be privatised away, how could this be done? This is an important entry point for understanding how racism could become so mainstream in the 2010s. Rather than claiming that the poor were just another neoliberal project away from becoming happy entrepreneurs, people were now increasingly attributing poverty to the background, or culture, or 'race' of a group, or saying that they lived in the wrong place, or that they were simply too lazy. Poverty was the fault of the poor. Racism became important in making people's worldviews fit together when neoliberalism no longer had anything to offer.

At the height of neoliberal hegemony, racism could almost appear as a special interest of the far right. When progressive neoliberals – to cite Nancy Fraser – talked about diversity and liberal immigration policies,

36 Perhaps we can add that the world's fools seem to be disproportionately resident in Sweden. Here, a hospital *was privatised* in the middle of the pandemic.

37 Larry Elliott, 'Biden Attempts to Consign Trickle-Down Economics to the Dustbin of History', *Guardian*, 29 April 2021.

one could almost get the impression that the ruling class and political elite were no longer interested in national chauvinism or racism.[38] All such illusions disappeared during that useless decade. The 2008 crisis challenged liberal pipe dreams of a borderless world. With the corona crisis in 2020 they collapsed hard and brutally.

During relatively strong hegemonies, the spheres of ideology, politics and economy interact and mutually reinforce each other. In crisis, this is not necessarily the case. The political project during the 2010s was to do whatever possible *not* to change the economy in general and the existing class relations in particular. If we return to levels of abstraction in chapter 3, we can say that this was crisis management merely at abstraction level one. Policies of lower interest rates, cheaper money and more debt were implemented so that level two – the actual organisation of neoliberalism – would not change. But underlying problems remained unresolved and the economic crisis needed a more thorough creative destruction. Concurrently, neoliberal ideology collapsed. Economic and ideological developments shook our world, while politicians tried desperately to keep alive what was already dead. The 2010s are an interesting example of what happens when rulers under capitalism try to avoid changes in a changing world.

We Only Had a Decade (Environmental Arguments)

Just weeks after Lehman Brothers went bankrupt in 2008, Deutsche Bank came on the scene and tried to describe the crisis as an opportunity for green investment. Similar analyses – or should we say hopes – came from the World Bank, the IMF, the International Energy Council and the UN Environment Programme.[39] After Al Gore's 2006 film *An Inconvenient Truth*, many leading politicians and the mainstream media seemed to have woken up. Gore and the IPCC were awarded the Nobel Peace Prize in 2007, and, by the end of 2009, up to 100,000 people were demonstrating at the UN Climate Change Conference (COP 15) in

38 For discussion, see Nancy Fraser and Rahel Jaeggi, *Capitalism: A Conversation in Critical Theory* (Cambridge: Polity, 2018), pp. 193–215; Anderson, 'Situationism à L'envers', pp. 89–90.

39 For an overview, see Joel Wainwright and Geoff Mann, *Climate Leviathan: A Political Theory of Our Planetary Future* (London: Verso, 2020), p. 109.

Copenhagen. There was historic pressure to stop climate change. Demonstrators knew that the 2010s would be a crucial decade to reverse the trend. At the Copenhagen summit, Bolivia's negotiator, Angélica Navarro Llanos, sought to mobilise for the radical change that was needed. 'We only have a decade' was her message to other leaders.[40]

It was a monumental failure. From 2010 to 2019, global emissions increased from 33.15 million tonnes of carbon dioxide to 36.57 million tonnes per year. The global average temperature rose from 0.91 to 1.15 degrees compared to pre-industrial times. If emissions had started to be reduced in 2010, according to the UN Environment Programme, emissions would 'only' need to be reduced by 3.3 per cent each year to limit warming to 1.5 degrees. Because the 2010s were so worthless, in 2019 we would need to reduce them by 7.6 per cent each year to reach the same target.[41] The decade was so useless that even peak oil theory did not work.

At the 2015 UN Climate Change Conference in Paris, the world's politicians succeeded in one thing: raising ambition. While all developments pointed well past 2 degrees, the target was changed from a maximum of 2 degrees to a maximum of 1.5 degrees. France's then president François Hollande called it 'a major leap for mankind', and David Cameron argued that the elites had now 'secured the planet for many, many generations to come'. Renowned climate scientist James Hansen called it a fraud. Countries were supposed to self-report their plans for what they voluntarily intended to do. Even if everyone actually did everything they said they could potentially do, climate scientists say there would still be a 90 per cent chance of temperatures rising by more than two degrees by 2100 and a 33 per cent chance of them rising by more than three degrees.[42] There were no obligations, and no mechanisms to punish countries that did nothing at all. All attempts at binding agreements failed.

---

40 Naomi Klein, *On Fire: The (Burning) Case for a Green New Deal* (London: Allen Lane, 2019), p. 28.

41 UNEP, *Emissions Gap Report 2019* (Nairobi: United Nations Environment, 2019), p. xx; IPCC, *Global Warming of 1.5°C* (Cambridge: Cambridge University Press, 2018).

42 Ian Angus, *Facing the Anthropocene: Fossil Capitalism and the Crisis of the Earth System* (New York: Monthly Review Press, 2016), pp. 89, 160–1, 167. For quotes from Hollande and Cameron, see Wainwright and Mann, *Climate Leviathan*, p. 35. For Hansen, see Oliver Milman, 'James Hansen, Father of Climate Change Awareness, Calls Paris Talks "a Fraud"', *Guardian*, 12 December 2015.

Climate policy in the 2010s was a period of endless optimism for 'green business'. Everything could be solved with money and markets. At an individual level, rich people can compensate for climate change, while at a state level, oil nations like Norway can continue fossil fuel production by participating in carbon emission trading or paying countries in the Global South for not cutting their forests.[43]

It is more difficult to discuss climate policy than economic policy in terms of neoliberalism, because it has never been dealt with under any other form of capitalism. Nonetheless, towards the end of the 2010s, non-neoliberal alternatives became increasingly clear. The strongest was from right-wing climate deniers. Another originated from the environmental movement, with Fridays for Future, Extinction Rebellion and spectacular actions by Ende Gelände. The energy in the movement from 2009 seemed to be back.

Both the ecological and economic crises were met in the 2010s with a series of non-solutions – always at the most superficial level of abstraction. With the economy, we were given more painkillers (debt, cheap money, lower interest rates); with climate policy, we were constantly given new promises (we'll solve it 'before 2030' or 'before 2050'). In both cases, it was about living beyond our means. Economic problems can be pushed ahead into unemployment, poverty and new crises; ecological problems can be pushed ahead all the way into death.

The then president of the UN General Assembly, María Fernanda Espinosa Garcés, argued in 2019 that we only had eleven years to avoid climate catastrophe. Once again, we seem to be in a crucial decade.[44] I remember thinking on New Year's Eve 2019 that it would be good to leave behind such a useless decade as the 2010s. I had just signed the contract for this book, and although I did not expect the crises to disappear, the coming decade could not be worse than the last. Surely it could only get better? Then the 2020s started.

43 See, e.g., Eivind Trædal, *Det svarte skiftet* (Oslo: Cappelen Damm, 2018), p. 128; Erik Martiniussen, *Drivhuseffekten: klimapolitikken som forsvant* (Oslo: Manifest, 2013), p. 145.

44 UN, 'Meetings Coverage ga/12131', 28 March 2019, General Assembly, seventy-third session, high-level meeting on climate and sustainable development.

The Corona Crisis as Social Paroxysm

The corona pandemic became an archetypical example of how easy it is to forget all previous signs of weakness and instability in the economy as soon as a crisis actually hits. The world was in shock, and it was all the fault of an extraordinary virus.

During 2020 it was often repeated that 'we cannot go back to normal'. *This* changes everything! The world of weekend flights to Paris and increasing inequality had apparently come to an end. Now we *had* to establish new relations to nature. This was true in one way: there is no going back when capitalism constantly transforms the world through crises. But we could indeed go 'back' to escalating climate change, increasing economic inequality and a highly unstable economy. Much of the naïveté around this being a moment – or opportunity – for things to change for the better came because the crisis was once again considered something extraordinary. And not, as it turned out, yet another societal paroxysm that reproduced capitalism. The so-called corona crisis must therefore be analysed at different levels of abstraction, and when using the five-level framework from chapter 3 it can be read as follows.

Level One: The Surface, the Event, the Shock

All crises are shaped by the factors that trigger them, but shocks can also be spellbinding. The shape of the 2020 crisis was certainly something out of the ordinary. Social and economic life came to a standstill. Borders were closed, and many shops, schools and universities, cinemas and nightclubs were closed to varying degrees in different countries to stop the virus from spreading. Some countries imposed curfews. In February and March 2020, the crisis entered the classic panic phase, with stock markets plummeting.[45] The pandemic was like a handbrake on production, circulation and consumption – simultaneously. In June 2020, we could read that some 50 million people were at risk of being pushed into extreme poverty, but the figure would turn out to be nearly twice that. According to the ILO, 8.8 per cent of all global working hours were lost that year, equivalent to 225 million jobs. Some have estimated that the

45 For data, see, e.g., Toussaint, 'The Capitalist Pandemic'.

economic crisis in 2020 was at least double the size of the 2008 crisis.[46] In addition, there is the pandemic itself, with millions killed by the virus.

The proposed solution was confined to abstraction level one: a simple vaccine against the virus would also be a vaccine against the economic crisis. Pandemic management was very much about enduring the problems – flattening the curve – while waiting for the great rescue: Big Pharma offering to sell us a vaccine.

The extraordinary shock was used by politicians who wanted to appear as strong leaders. However, striking a balance between saving the economy and stopping the spread of Covid-19 was not straightforward, especially when the contagion returned with mutations and new waves. The challenge for the politicians was to simultaneously keep capitalism and its own population alive, and this was done by shutting down parts of the economy with one hand and providing massive subsidies and cheap money with the other.

State intervention in people's lives was necessary from a health perspective. At the same time, there is no escaping the fact that the corona crisis could have been the perfect pretext for politicians to curtail rights and perhaps even approach outright fascism. Even if Giorgio Agamben was far out of line when he argued against any kind of restrictions, it is certainly true that massive shutdowns and harsh restrictions on people's freedom potentially *could* be used by highly reactionary forces.[47] This never became a dominant trend as many right-wingers were more preoccupied with conspiracy theories and driving demonstrations against lockdowns and vaccines.

In the context of discourse, the cause (virus) and the solution (vaccine) to the crisis remained on the surface. The crisis itself, of course, went much deeper.

---

46 On extreme poverty, see IMF, *World Economic Outlook: Managing Divergent Recoveries* (Washington, DC: IMF, April 2021), p. xvi; Daniel Mahler et al., 'COVID-19 Could Push 100 Million People into Extreme Poverty, Says World Bank', World Economic Forum, weforum.org, 12 June 2020. On ILO, see *ILO Monitor: COVID-19 and the World of Work* (Geneva: International Labour Organization, 2021), pp. 1, 9. On 2020 and 2008, see also Ashley Smith and Michael Roberts, 'After the Pandemic Slump, What Next? Interview with Michael Roberts', *Spectre*, 25 February 2021.

47 Giorgio Agamben, 'The Invention of an Epidemic', *European Journal of Psychoanalysis*, journal-psychoanalysis.eu, 26 February 2020.

Level Two: The Concrete Organisation of Capitalism

The enormous shutdowns that followed the pandemic would clearly have affected the economy, no matter how it was organised. But, if a similar pandemic had occurred in the 1950s or 1980s, it would have looked very different. In the 1950s, politicians could have faced the crisis with large investments, and in the 1980s within the framework of emerging and positive neoliberalism. Now, the pandemic faced societies that had experienced decades of neoliberalism and were in the middle of an organic and hegemonic crisis. The infrastructure and government systems that were supposed to deal with social disasters had been weakened by cuts, privatisation and lack of investment. In Sweden, for example, 96 per cent of all municipalities planned to cut back on care for the elderly in 2020 and emergency stocks had been abolished. The number of intensive care units equipped with ventilators decreased dramatically over twenty-seven years, dropping from 4,300 to fewer than 600.[48] Such developments were always criticised from various quarters, and the World Health Organization outraged many conservatives and liberals when it concluded in 2008 that *inequality kills* on a massive scale.[49] The pandemic certainly came at a bad time.

The situation called for extraordinary measures, and according to some figures, the states' crisis packages in 2020 together totalled as much as $14 trillion – significantly higher than in 2008.[50] The neoliberal ideology of austerity was blown away as governments continued to pump cheaper money into the system, with more qualitative easing and low interest rates. It was again *whatever it takes* to save the system. That the extraordinary injection of money did not lead to higher repo rates shows that the economy was not healthy; that the economy was likely to crash *if* repo rates were raised shows a crisis in crisis management.

The pandemic was discussed in 2020 from every possible angle,

48 Roya Hakimnia and Ståle Holgersen, 'Corona, hälsa och klass. Klassen som skapat krisen är inte den som får betala', *Röda Rummet* 187–8 (2020), pp. 3–6.

49 The Spanish professor of public health Vincente Navarro argues that it is not inequality per se that kills. It is *the people* who benefit from inequality that kill other people. See Vicente Navarro, 'What We Mean by Social Determinants of Health', *International Journal of Health Services* 39, no. 3 (2009), p. 440.

50 Tooze, *Shutdown*, p. 131. This came with wide geographical disparities, see ibid., chs 5–8.

except the most important one: How to avoid future pandemics? Pandemic management was basically reduced to social distancing and vaccines, but there were exceptions. The United Nations Environment Programme (UNEP) has proposed One Health, an approach that brings together public health, veterinary and environmental expertise to prevent and respond to outbreaks of zoonotic diseases and pandemics.[51] Björn Olsen, professor of infectious diseases, argued there should be about twenty surveillance systems per continent in different niches where ecologists, biologists, veterinarians and others work together and monitor problems such as emerging influenza, coronaviruses and other viruses.[52] Given that this is a purely intra-capitalist solution, it is curious that such discussions were almost exclusively among scientists and activists.

Why was it so difficult for those in power to discuss the underlying causes of the pandemic? In part, the commercialisation of nature and the destruction of wilderness are highly sensitive issues. Carbon emissions were a known problem decades before any action was taken to reduce them (if we are even there). How many more pandemics will it take for us to really dare to discuss these problems? Perhaps it is relevant that scientists predict that future zoonotic diseases will often come from Africa: Will future defences include a combination of vaccines, walls and moats? That economic crises bring change is something that politicians and capitalists have learned to live with – and often exploit. Our relationship with nature is another matter. Compared to how profitable it is to destroy it, there are only crumbs to be made by saving it.

Level Three: Inherent in Capitalism

The One Health perspective that exists within UNEP would be light years better than completely ignoring the issue, but this approach also ignores the underlying causes of the pandemic. Rob Wallace and his colleagues have therefore proposed Structural One Health to emphasise

51 UNEP, *Unite Human, Animal and Environmental Health to Prevent the Next Pandemic*, United Nations Environment Programme, press release, 6 July 2020.

52 See Christer Bark, 'Björn Olsen om den bristande beredskapen, framtiden och vad som behöver göras', *Sjukhusläkaren*, 18 April 2020.

the processes that cause zoonoses. One such approach points to the metabolic rift and capital accumulation.[53]

If we really want to stop future ecological crises, we need to start healing the metabolic rift – which is not a new idea. Engels concluded in 1872 that humans should return to the earth what they have received from it.[54] After 150 years of non-renewable fuels, the idea of a sustainable metabolism has become almost bizarre. The fact that this sounds so alien today exemplifies the importance of the metabolic rift.

In the 2010s, we saw what can happen when politicians try to prevent economic crises from being resolved through creative destruction. The corona crisis of 2020 was yet another missed opportunity in this respect. Overproduction and the increasing organic composition of capital never had to face their main counter-tendencies: destruction and devaluation, new technological revolutions and radical reorganisation of capitalism. Previous analyses of the economic consequences of pandemics show that it takes a long time for the economy to recover: pandemics are not like wars where the physical environment has to be rebuilt immediately.[55] The pandemic deepened the crisis of neoliberalism.

Levels Four and Five: Endless Pursuit of Profit; Use-Value and Exchange-Value

That the profit motive can be one of the causes of the crises is unimaginable for the defenders of capitalism. The ruling class has built its world on capital accumulation; therefore it will never admit that the profit motive has a Janus face, that it creates both booms and economic crises; that it can bring both 'green growth' and climate change. It is impossible to imagine that the cure for crises has also caused them.

53 See Rob Wallace et al., 'The Dawn of Structural One Health: A New Science Tracking Disease Emergence along Circuits of Capital', *Social Science and Medicine* 129 (2015), pp. 68–77.

54 Friedrich Engels, *The Housing Question* (London: Martin Lawrence, 1942), p. 95.

55 John Authers, 'When Plagues Pass, Labor Gets the Upper Hand', Bloomberg, 5 April 2020. For discussions, see also Òscar Jordà, Sanjay R. Singh and Alan M. Taylor, *Longer-Run Economic Consequences of Pandemics*, Working Paper No. 26934, National Bureau of Economic Research, 2020; Olga Khazan, 'How the Coronavirus Could Create a New Working Class', *Atlantic*, 15 April 2020.

Covid-19 was caused by the profit motive. Small farmers in China pushed closer to the wild with urbanisation, deforestation and industrial agriculture and the subsequent amped-up hunt for rare animal species brought to markets – the place from which the virus most likely originated. To minimise the risk of future pandemics, we must stop destroying wilderness, but also confront capital accumulation. One intra-capitalist solution to zoonoses and new pandemics points in a different direction, namely more human control over nature and, its logical consequence, eradicating all wilderness and wildlife.[56]

The coronavirus pandemic was rooted in the same underlying contradiction as other capitalist crises: between use-value and exchange-value, and between the abstract and the concrete. Animals will still be animals, no matter how much they are also treated as abstract capital. The swine flu (H1N1) pandemic of 2009 came from pig farms, places where an owner can coldly calculate that each pig might get 1 or 1.5 square metres to live on, and that the pigs will yield the most profit if they are allowed to live for four or six months. Treating animals as abstract commodities creates concrete crises. No market, no investment, no efficiency or rationalisation can conjure nature away: exchange-value needs its use-value.

The pandemic made it particularly clear what all capitalist crises are ultimately about: death. Health has always been strongly linked to class, and with Covid-19, (un)health was placed in the middle of the class struggle. Several Marxists fruitfully discussed the crisis in terms of *capital versus life*.[57]

When 'nature' kills, it never comes alone. Death is as social as the life we live. And it comes within political systems.[58] Here, different levels of abstraction are linked together. *Capital against life* is an abstract formulation that reflects capitalist crises in general, but the pandemic demonstrated this on a concrete level. On the first level of abstraction, we find direct death. In previous crises, death has often been the consequence of

56 See also Andreas Malm, *Corona, Climate, Chronic Emergency: War Communism in the Twenty-First Century* (London: Verso, 2020).

57 Marxist Feminist Collective, 'On Social Reproduction and the Covid-19 Pandemic: Seven Theses', *Spectre*, 3 April 2020.

58 Rob Wallace, *Big Farms Make Big Flu: Dispatches on Influenza, Agribusiness, and the Nature of Science* (New York: Monthly Review Press, 2016); Mike Davis, *The Monster at Our Door: The Global Threat of Avian Flu* (New York: New Press, 2005). For a broader discussion of health and class, see, e.g., the work of Howard Waitzkin, Vicente Navarro, Göran Dahlgren and Friedrich Engels.

the crisis. This time it triggered the crisis. Death was the shock itself. At abstraction level two, we find a neoliberal system that for decades has underinvested in, or privatised away, vital infrastructure. At abstraction level three, it is clear that we are still, to paraphrase Engels, taking something from the planet and giving something else back. When we take out fossil fuels and return toxins, it is not only we who are killing nature; the metabolic rift is also killing us.

The Great Equaliser? Corona and Class

One infamous moment in 2020 was when pop legend Madonna, with a fortune of $850 million, sat naked in a bathtub in a viral clip and talked about coronavirus as 'the great equalizer'.[59] Magnate David Geffen isolated himself on his $590 million yacht, *Rising Sun*, and posted a picture of the yacht at sunset on Instagram, with the caption 'Hope everyone is safe'. The corona pandemic began with a tenfold increase in private jet flights in the US. In the UK, there was a torrent of requests for mansions. Self-isolation had different geographical meanings for the very rich than for the rest of us. London's richest were willing to pay £10,000 a week in rent for luxury villas and mansions, now that they had to have 'home offices'.[60]

In Europe, Covid-19 first spread via relatively privileged ski tourists in the Alps, and in Brazil the crisis was called 'the disease of the rich' because, in the early phase, many wealthy people were affected. Donald Trump and Prince Charles contracted the disease, and Boris Johnson was in intensive care. It is in this light that we must view Madonna's strange video. Many privileged people genuinely feared for their lives.

Then it turned out that Covid-19 had the same class character as other ecological crises. In Rio de Janeiro, sixty-three-year-old maid Cleonice Gonçalves was infected by her employer, who had been on holiday in Italy. On the same day that *he* received a positive result on his test, *she* died. In England, Boris Johnson woke from a coma and urged anyone unable to work from home to return to their jobs, which, as Owen Jones pointed out, meant factory workers, but not managers;

59 Toyin Owoseje, 'Coronavirus Is "the Great Equalizer", Madonna Tells Fans from Her Bathtub', CNN, 23 March 2020.

60 See Malm, *Corona, Climate*, p. 92.

cleaners, but not accountants. The corona crisis was a new classquake. Even in this crisis, we were not all in the same boat. Covid-19 was by no means a leveller. And so, too, Madonna was forced to delete her post and Geffen deleted his entire account on Instagram.[61]

For some, the crisis was an opportunity. Some figures border on the absurd. In nine weeks in 2020, Jeff Bezos earned $334 billion while 38 million Americans lost their jobs.[62] Many individual capitalists lost a lot, of course, but the crisis hit differently depending on sectors, class fractions, organisational form, nationality and so on. Between the onset of the crisis and January 2021, the wealth of ten people increased by an amount that could prevent the entire population of the planet from falling into poverty due to the virus *and* pay for vaccines for all the people on our planet.[63] A defender of capitalism might say that, even if the figures are correct, this has nothing to do with class: it would not help all the world's poor if the ten richest did not make astronomical amounts of money – these are two different worlds. And, in some respect, they *are* indeed two different worlds. Class is precisely the reason why these are simultaneously different and interconnected worlds.

Crisis management in 2020 came with a familiar class character. To avoid total collapse, governments launched gigantic aid packages. Some of them were targeted more at the working class than others, but the main project was yet again about saving capital. High profits are often justified with the argument that capitalists take risks and therefore deserve a little extra when things go well. By the same logic, they would then inject capital when things go badly. Instead, many received state aid. One example that reached the news in Sweden was when SKF (originally Svenska Kullagerfabriken) received state support in the spring of

---

61 On Gonçalves, see Yngvild Gotaas Torvik, 'Fra Madonna Til Rinkeby', *Klassekampen*, 20 May 2020, pp. 18–19; Owen Jones, 'Boris Johnson's Message to the Working Class: Good Luck Out There', *Guardian*, 12 May 2020. See also Oxfam's crisis report, aptly titled *The Inequality Virus*: Esmé Berkhout et al., *The Inequality Virus: Bringing Together a World Torn Apart by Coronavirus through a Fair, Just and Sustainable Economy* (Oxford: Oxfam International, 2021), pp. 2, 28. On Madonna, see Pål Velo, 'Formuene eser ut i krisa', *Klassekampen*, 6 May 2020.

62 Sarah Ruiz-Grossman, 'Billionaires' Net Worth Rises during Pandemic as Millions Go Unemployed: Report', *Huffington Post*, 22 May 2020. The figure on new unemployed does not include migrant workers and others who are unable to register.

63 Berkhout et al., *The Inequality Virus*, p. 2.

2020, while distributing 120 million euros to its shareholders.[64] The Danish decision that companies registered in tax havens would not receive crisis aid attracted much international attention, although it turned out that such a rule was difficult to put into practice.[65] The point under debate was whether the states should give money to companies that, at the same time, distribute bonuses and profits to owners who, in turn, place the money in tax havens. This is class power during a crisis. What is considered good for the capitalist class is considered good for all of us. The ruling class with old money and pedigrees is considered responsible and forward-looking, while the ruling class with new money is considered young, innovative and liberal. It seemed politically impossible to imagine crisis solutions beyond their class interests.

All over the world, workers, the poor, the racialised, people in prison, refugees and undocumented people were hit hardest.[66] In context of stratification, the general trend was that the poorer you were, the harder you were hit. Mike Davis argued that, from a global perspective, we had two pandemics and two humanities: one humanity that was fed and had access to health care, and another that was hungry and lacked health care.[67] Two billion people in the Global South are constantly exposed to hunger and a range of infectious diseases. Clean water and soap exist for only 15 per cent of the population in sub-Saharan Africa. Kenya has 130 ICU beds and 200 intensive care nurses. In war-torn Yemen, Covid cases increased as health services collapsed; in the Gaza Strip, vaccines were missing and the only Covid-19 testing station had been bombed by Israel. In India, car parks and public parks were turned into cremation grounds.[68]

64 Kristina Lagerström and Johan Zachrisson Winberg, 'Large Companies Give Billions to Shareholders – Still Want Government Support', SVT, 4 May 2020.

65 Ingrid Hjertaker, 'Kan krisepakkene gjøres skatteparadisfrie?', *Agenda Magasin*, 24 April 2020.

66 For the US, see Ed Pilkington, 'Black Americans Dying of Covid-19 at Three Times the Rate of White People', *Guardian*, 20 May 2020; Kim Parker, Juliana Menasce Horowitz and Anna Brown, 'About Half of Lower-Income Americans Report Household Job or Wage Loss Due to COVID-19', pewresearch.org, 21 April 2020. For Sweden, see Dagens Nyheter, 'Så fick svensksomalierna skulden för sin egen död', *Dagens Nyheter*, 26 May 2020.

67 Mike Davis, 'Covid-19 Is the Biological Earthquake That Science Has Been Warning about for Almost a Generation', *Progressive International*, 30 April 2020.

68 Ibid.; see also Arundhati Roy, 'We Are Witnessing a Crime against Humanity', *Guardian*, 28 April 2021.

In the Global North, too, the poorest were hardest hit. They had less opportunity for social distancing; they drive less so it was harder to avoid public transport; they live in overcrowded conditions making it harder to isolate when ill, and so on. If sick leave means losing important income and perhaps even your job, you still go to work even if you are sick. Next, it is about vulnerability: the poorest generally have worse health, with more chronic illnesses in lower sections of the working class.[69] From a Weberian class perspective – which emphasises market positions – we can see how different jobs were affected differently. Workers in high-contact occupations are more exposed to the virus. Over 600,000 Swedes had private health insurance policies that they were able to take advantage of even in the midst of a burning pandemic. The fact that health care had largely become a market took on grotesque implications. In Stockholm in 2020, when the public sector was forced to deal with the pandemic with seventeen-hour shifts in intensive care, and while mortality rates were shockingly high, the private sector continued with its cosmetic surgery.

The fact that the right was generally less receptive than the left to restrictions and lockdowns – with Sweden to some extent an international exception – must be understood partly as relating to ideology and partly as a direct result of class interest. The right's ideological opposition to restrictions ties in with far-right conspiracy theories and religious fanaticism (as can be illustrated by Bolsonaro's Brazil), but, perhaps more importantly, it is motivated by pure class interest (as can be illustrated by news of workers dying in Covid outbreaks in meat plants in the US that, despite the social shutdown, kept open to secure their profits).[70] In Sweden, the state and employers downplayed material shortages and lack of resources in the health care sector, and instead invoked questionable science to justify reduced requirements for personal protective equipment.[71]

69 Mahler et al., 'COVID-19 Could Push'. One perhaps unexpected effect of the lockdown was that Japan had 20 per cent *fewer* suicides compared to the previous year, as people in the early phase of the crisis worked less, commuted less and had more time for family. See Gavin Blair, 'Japan Suicides Decline as Covid-19 Lockdown Causes Shift in Stress Factors', *Guardian*, 14 May 2020.

70 Anders Tvegård, 'President Bolsonaro mener koronakrisen er lureri fra mediene', *Nrk*, 24 March 2020; Daniel Werst, '800 Workers at a Pork Plant Get Covid-19: The Company Is Responsible', *Left Voice*, 28 April 2020.

71 Mikael Grill Pettersson, 'SVT avslöjar: Kommunernas intresseorganisation fick myndigheter att tona ned munskyddskrav', SVT, 26 April 2020.

If we remain only within a Weberian analysis, it is easy to miss the question of ownership, including the concept that someone actually owns health care.[72] As health care becomes more and more commodified, we see that clinics and pharmacies in Sweden are closing in poorer areas and opening in richer ones. When the pharmaceutical industry is driven by the profit motive, vaccines are restricted by patent – this morbid expression of the dominance of exchange-value over use-value. The tensions between the profits for big companies in one of capitalism's biggest sectors (AstraZeneca, Pfizer, Moderna) and the economy at large, which would arguably benefit from free vaccines, will probably be resolved somewhere in a showdown in the corridors of power. Today's pharmaceutical companies only become interested when there is a demand, i.e., when the disaster is a fact, and are not focused on preparing for the next possible pandemic by proactively developing vaccines. Eric Cazdyn argues similarly, in *The Already Dead*, that the capitalist pharmaceutical industry is not inclined to develop medicines that cure diseases; rather, it helps us live as the chronically ill.[73]

The profit motive ties the crisis together at different levels, from Chinese agribusiness, wild food production and the displacement of small farmers, to health and social care being increasingly in the hands of private owners; from the power of Big Food and Big Pharma, to state crisis management that is about compensating the ruling class for lacking profits. This is a system driven by the capitalist class and steered after their interests. Remember that it was not as high-income earners or highly educated or creative people that the world's richest in 2020 became wealthier to the extreme – it was as capitalists.

Combine stratification (the poorest have been hardest hit), Weberian analysis (which professions had home offices) and Marxist insights (the profit motive ties the crisis together), and the general conclusion ends up in a familiar place. *The class that caused the corona crisis was not the class that paid the price.*

72 See Roya Hakimnia, 'Att förstå hälsa genom klass och klass genom hälsa', *Fronesis*, 30 March 2021.

73 Eric Cazdyn, *The Already Dead: The New Time of Politics, Culture, and Illness* (Durham, NC: Duke University Press, 2012).

The Climate in the 2020s

There was one crisis that 2020 was not about. The climate crisis tends to disappear from the political agenda as soon as the economy starts to struggle, and 2020 was certainly no exception. The climate crisis itself, however, did not disappear.

In Europe's recorded history, 2020 was the warmest year. Some 9.8 million people had to flee what are known as hydrometeorological hazards and disasters in the first half of 2020 alone: Cyclone Amphan hit eastern India and Bangladesh; giant typhoons devastated the Philippines; and in Jakarta alone 60,000 were evacuated due to flooding. The average temperature in the Arctic was six degrees warmer than the 1981–2010 reference period; parts of South America suffered extreme drought; the US experienced the largest forest fires on record; Australia broke heat records; and giant locust swarms in East Africa and South Asia threatened the food security of tens of millions of people.[74]

Yet the exact same thing happened in 2020 as in 2008, and it would happen again in 2022 with the so-called energy crisis and inflation: recovery efforts were all about restoring growth and saving profits. It was crystal clear what the priorities were and what could simply wait.

Yet – despite less political discussion, despite the climate movement's lost momentum with the lockdowns, and despite the fact that 66 per cent of the 2020 recovery package in a country like Sweden went to measures that promote fossil energy – 2020 was an exceptionally good year for reducing global warming.[75] As countries began to shut down, carbon dioxide emissions plummeted. It was the biggest drop ever in absolute terms (about 2 billion tonnes of carbon dioxide), and the biggest since the Second World War in relative terms (5.8 per cent).[76]

---

74 See, e.g., WMO, *State of the Global Climate 2020*, GWMO-No. 1264 (Geneva: World Meteorological Organization, 2021); Belén Weckström, 'År 2020 var världens varmaste – Europa har aldrig sett något liknande', svenska.yle.fi, 8 January 2021; Kaamil Ahmed, 'Locust Crisis Poses a Danger to Millions, Forecasters Warn', *Guardian*, 20 March 2020.

75 Ylva Rylander and Gregor Vulturius, 'Stor del av Sveriges återhämtningspaket för COVID-19 har hittills satsats på åtgärder som främjar fossil energi', Stockholm Environment Institute, 9 November 2020.

76 IEA, *Global Energy Review: CO_2 Emissions in 2020*, International Energy Agency Report, iea.org, 2 March 2021; see also Dan Boscov-Ellen, 'Infectious Optimism: Notes on Covid-19 and Climate Change', *Spectre*, 27 May 2020.

Fewer cars on the road improved the local environment; in big cities the air was breathable and people in megacities could see the sky, which used to be hidden by exhaust fumes and smog. From Venice it was reported that, without giant cruise ships and tourists, the water was finally blue and clear.[77]

There was no need to open the champagne. The extreme figure of 5.8 per cent reduction in CO_2 emissions is almost 2 per cent *less* than the 7.6 needed each year to meet the target of 1.5 degrees. Everything was done to save capitalism, and no one was surprised when emissions increased dramatically the following year. The increase in 2021 was the largest annual increase ever, even larger in absolute terms than the decline the year before. The 6 per cent increase in emissions in 2021 is, of course, closely related to the fact that the economy grew by 5.9 per cent in the same year.[78]

The 2020 pandemic offers at least three lessons concerning the climate issue. First, the most important thing is not to reduce individual consumption. There are arguments that the pandemic reduced simple emissions; for instance, people stopped flying as frequently. Reducing the *next* 7 per cent, and then the next 7 and the next 7, requires quite different policy regimes than lockdowns and home offices. To avoid more than 1.5 degrees of global warming, we need a radical restructuring of the world economy, which again requires rock-hard government regulation of the capitalist class. Capital should not be motivated or offered rewards; it must be controlled or expropriated. Secondly, the radical measures put in place to stop Covid-19 show us something we really already knew, namely that political power carries enormous possibilities and that states can adjust to major challenges. Third, 2020 gave us a profound insight into the diabolical crisis: politicians who accept the capitalist imperative do not have to solve the climate crisis, but they do have to get capitalism back on its feet. Economic crises *must* be solved – at any cost. The climate crisis must not.

77 John Brunton, '"Nature Is Taking Back Venice": Wildlife Returns to Tourist-Free City', *Guardian*, 20 March 2020.

78 IEA, *Global Energy Review: CO_2 Emissions in 2021: Global Emissions Rebound Sharply to Highest Ever Level*, International Energy Agency Report, iea.org, 2022, p. 3.

The Crisis of Neoliberalism, or: Thoughts on How the Crisis Reproduces Capitalism This Time

Covid-19 did what politicians were desperately trying to avoid: it ended neoliberalism. I truly think this is the case, but the statement is anything but straightforward. Since 2008, countless texts have been published on the (non-)end of neoliberalism, but what does it mean for something like neoliberalism to end? This is difficult to answer, since people define the term differently, or hardly define it at all.[79] The advantage with the concept neoliberalism was always also its weakness. Its broad scope managed to tie together elements that were related, but made the concept ambiguous and often impractical for analysing concrete events.

In the decades after the Second World War, different concepts were used to analyse production (e.g., Fordism), fiscal and monetary policy (Keynesianism), dominant political movements (social democracy) and ideological beliefs (modernism). Today, all this and much more are included in one concept. Neoliberalism often appears as a bad abstraction. Defined broadly enough, it can survive anything. For David Harvey, for example, neoliberalism is a class project.[80] This is true. But it cannot be *any* class project in which the capitalist class gains more power over workers. Most authors agree that states were always central under neoliberalism, but not *all* state policies can be neoliberal. If

79 On neoliberalism as an academic 'buzzword', see Taylor Boas and Jordan Gans-Morse, 'Neoliberalism: From New Liberal Philosophy to Anti-liberal Slogan', *Studies in Comparative International Development* 44, no. 2 (2009), pp. 137–61; James Ferguson, 'The Uses of Neoliberalism', *Antipode* 41, no. S1 (2010), pp. 166–84. On neoliberalism, see, e.g., David Harvey, *A Brief History of Neoliberalism* (Oxford: Oxford University Press, 2007); David McNally, 'Slump, Austerity, and Resistance', in Leo Panitch, Gregory Albo and Vivek Chibber (eds), *Socialist Register 2012: The Crisis and the Left* (London: Merlin, 2011), pp. 36–63; Gérard Duménil and Dominique Lévy, *The Crisis of Neoliberalism* (Cambridge, MA: Harvard University Press, 2011); Ståle Holgersen, *Staden och kapitalet. Malmö i krisernas tid* (Gothenburg: Daidalos, 2017); Jamie Peck, Nik Theodore and Neil Brenner, 'Neoliberal Urbanism Redux?', *International Journal of Urban and Regional Research* 37, no. 3 (2013), pp. 1091–9; Wendy Brown, *Undoing the Demos: Neoliberalism's Stealth Revolution* (Cambridge, MA: MIT Press, 2015); Ben Fine and Alfredo Saad-Filho, 'Thirteen Things You Need to Know about Neoliberalism', *Critical Sociology* 43, no. 4–5 (2017), pp. 685–706.

80 See David Harvey, *The New Imperialism* (Oxford: Oxford University Press, 2003); Harvey, *A Brief History*.

fascism is neoliberal and all traces of liberalism disappear from neoliberalism, the question is whether we have any use for the concept at all.

Many left intellectual milieus have since the early 2000s been so deeply engaged in discussions about neoliberalism that it has become hard to think beyond it. Yanis Varoufakis has suggested that we are experiencing the end of capitalism and are now in the onset of 'techno-feudalism'.[81] Here comes the latest version of collapse theory: the crisis of neoliberalism misconceived as the final crisis of capitalism. To paraphrase Fredric Jameson, it seems easier to imagine the end of capitalism than the end of neoliberalism. It is more plausible to assume that the economic crisis will pass once more and new ways of organising capitalism will emerge. On a slightly warmer planet, the ruling class, with new-found optimism for the future, will once again proclaim that the time for crises is over.

A common mistake in the history of economic crises comes from drawing individual trends from the present into the future. It is perhaps inevitable that people emphasise the crisis they have just been through or are still going through when discussing the future. This is indeed a challenge. In the interwar period, working-class poverty was highly visible – even within crisis theories – and it was normal to think that capitalism would create more poverty and polarisation, not infrequently all the way to the collapse of capitalism. The book Habermas published in 1973 on why contradictions in capitalism did *not* cause economic crises, but only other types of crises, could not have been written a few months later. The crises of the 1970s, in contrast to those of the interwar period, were seen by few left-wing groups as the beginning of the end of capitalism. The relative strength of the working class, however, enabled crisis theories such as the *profit squeeze*, and also broader theories of Italian *operaismo* in the 1960s, which held that the active agency and strength of the working class had always been at the heart of capitalist development. Today, when the working class has been systematically harassed for decades, people's beliefs are more likely to go to the other extreme: it does not matter what the working class does, as nowadays everything happens because of capital, or perhaps 'financialisation'.

Since inequality has increased radically under neoliberalism, it is

81 Jason Myles and Pascal Robert, 'Yanis Varoufakis: We Are Living in a Post-capitalist, Techno-feudalist Dystopia', *Real News*, 22 February 2022.

tempting to draw such trends into the future as well. Up until 2022, it was common to ignore potential problems with inflation – which even opened a discursive space for Modern Monetary Theory – but, during 2022, inflation became *the* thing to handle. A renewed interest in the collapse of capitalism in our time is probably linked to the sense of doom that can be derived from climate science. Increasing private and public debt – within and between the Global North and South – is also a factor that could contribute to economic and political instability in the coming years. How this will play out, we do not know. States have, arguably, played a more prominent role since 2020 (or 2008) than we were used to under neoliberalism, with more state-owned enterprises and larger sovereign wealth funds; an increasing degree of economic nationalism; a renewed focus on state-directed investment in international development; and the increasing influence of the Chinese version of capitalism. Some go straight from such observations to wanting to talk about our times as *state capitalism* or *state monopoly capitalism*.[82]

We saw in the introduction that the young Marx was misled when he brought views on crises from the eighteenth century – when crises were often associated with revolutions – into his own century. Our challenge today is to learn from the twentieth century without being blinded. One striking phenomenon in the twentieth century was the way the Great Depression and Great Stagflation were resolved through very dramatic changes in capitalism. It is fairly easy to see how crises contributed to the beginning and the end of the Fordist-Keynesian era. Many authors – including myself – believed the 2008 crisis would end neoliberalism just as dramatically as crises changed the world in the 1930s and 1970s. Were we blinded by the twentieth century? Is it a mistake today to expect that neoliberalism should end through a similar dramatic change?

One argument that neoliberalism is not dead is that we still cannot define what system will come after it. The concept of neoliberalism was used to describe many different things, and many seem to be waiting for a new master-signifier that will replace neoliberalism more or less simultaneously in all spheres of life. But changes in, for example,

82 See Grace Blakeley, *The Corona Crash: How the Pandemic Will Change Capitalism* (London: Verso, 2020); Ilias Alami et al., 'Special Issue Introduction: What Is the New State Capitalism?', *Contemporary Politics* 28, no. 3 (2022), pp. 245–63.

ideology, politics and economy do not develop exactly the same way at the exact same tempo. Rather than searching for one new master concept that will define everything, I think it is more helpful to operate with different concepts to describe different things.

Are we misled by the twentieth century when we put too much emphasis on major hegemonic crises that seem to occur every forty to fifty years? Have not the crises of the last decades come with a temporality more similar to those of the nineteenth century, with crises approximately every nine to ten years (1973, 1982, 1991, 2000, 2008 and 2020)?[83] Crises every ten years can surely coexist with larger hegemonic crises every forty to fifty years, but is the former a more politically relevant starting point than the latter for understanding economic development during the twenty-first century? I certainly do not have any final answers here; I can only stress the need to avoid dogmatism.

When I insist that neoliberalism is dead, it is because I think it is a major intellectual mistake with political implications to use neoliberalism as a lens, a theory or an entry point to understand the 2020s. In terms of politics, organisation of the economy, and ideology, there have been such massive changes that the concept should be considered outdated. What makes the question a bit tricky is that the class power that constituted neoliberalism is still with us.

Class and the End of Neoliberalism

In chapter 3, I suggested that neoliberalism survived longer than the Fordist-Keynesian era partly because neoliberalism was more prone to crises. Now we must also include class in the equation. One reason the stagflation crisis in the 1970s was resolved relatively quickly was because powerful fractions within the capitalist class – that had police and military backing – were able to implement alternatives to post-war social democratic Keynesianism. The direction they pointed in was almost a given – smash the unions, cut wages, give freedom to capital.

Just over a decade after 1973, politicians were steering the political economy towards what became known as neoliberalism. Now, almost two decades after 2008, we still do not know exactly where we are going.

83 See also David McNally, *Global Slump: The Economics and Politics of Crisis and Resistance* (Oakland: PM Press, 2011), pp. 38–40.

Generally speaking, in the 1970s, strong factions of the ruling class wanted to end the existing regimes; in the 2010s the hegemonic factions of the ruling class *did not* want to end neoliberalism. Why end the neoliberal order from which they had benefitted so massively? Since 2008, neoliberalism has been falling apart, but the class power supporting it has not.

In contrast to the Great Depression, when even the richest lost out, the useless decade and the corona crisis of 2020 were periods when the richest became even richer. Large masses of people in countries such as Italy, Portugal, Greece, Spain and England were plunged into depression and poverty while the number of billionaires increased.

There were several factors associated with the inflation of 2022: a surge in energy prices due to the Russian war on Ukraine; massive state spending during the pandemic and then the re-opening of the economy after lockdowns; and supply chain problems, not least as China retained lockdowns. Mainstream economics often blame too much cheap money and quantitative easing between 2009 and 2019 as causes for the inflation, but Costas Lapavitsas, James Meadway and Doug Nicholls argue this cannot explain a *general* increase in prices because the financial institutions mainly used this money to trade in financial assets and real estate. Rather, one absolutely crucial factor was that firms raised prices to increase profits. Lapavitsas and colleagues show how the big businesses that dominate production and distribution in the UK profited massively from high inflation, while working people were losing: 'The source of record profits is the fall in real wages as inflation rises. To put it plainly, a large part of the income of working people is being transferred directly into the profits of big business.'[84] Even the IMF argued that 45 per cent of the price increase was caused by companies seeking higher profits. Both the European Central Bank and OECD made similar arguments.[85]

Attempts to stop inflation would, therefore, logically include regulating profits and implementing price controls. Due to the existing balance of class

84 Costas Lapavitsas, James Meadway and Doug Nicholls, *The Cost of Living Crisis (and How to Get Out of It)* (London: Verso, 2023), pp. 8, 38–45.

85 For IMF, see Niels-Jakob Hansen, Frederik Toscani and Jing Zhou, 'Europe's Inflation Outlook Depends on How Corporate Profits Absorb Wage Gains', imf.org, 26 June 2023; for ECB and OECD, see Isabella Weber, 'Taking Aim at Sellers' Inflation', socialeurope.eu, 18 July 2023.

forces, this just could not happen. The bill – yet again – had to be paid by the working class, which meant higher interest rates and real-wage reductions. And, again, we were told that we were all in the same boat, and due to war we really needed to unite our nations and support militarism. Meanwhile, we lived de facto with class struggle from above in a world on fire.

The balance of class forces has largely remained intact, with some temporal and geographical exceptions. This made it possible for the ruling class and leading politicians to experiment, despite an unhealthy economy, with new ways of organising the political economy. Some hoped that grand coalitions between social democrats and conservatives (e.g., Germany 2013–21) would stabilise things until the storm had passed. Some conservatives and liberals went into alliances with the far right (e.g., Sweden 2022–), while others converted into far-right populists themselves (e.g., UK 2019–22). Yet others have tried to go 'green', either with massive spending and subsidies (e.g., the Biden administration) or in political ambitions (e.g., the EU's green deal), hoping this will be economically and geopolitically advantageous. And some have taken a step to the left (e.g., Greece 2015–19), trying to recreate something that resembles a post-war social democracy. And then there was Liz Truss, the UK prime minister who wanted to test one last time if neoliberalism really was dead. In September 2022, she famously suggested abolishing the income tax rate of 45 per cent applied to those earning more than £150,000 a year. This typical neoliberal crisis management would certainly have been fuel to the fire concerning the ongoing inflation, and the proposal roiled financial markets and sent the British pound into a tailspin, and UK pension funds came close to collapse amid an unprecedented meltdown in UK government bond markets.[86]

In the mid-2020s, members of the ruling class are becoming increasingly aware that the old ways of organising the political economy – what we call neoliberalism – are outdated. And, for that reason, they are searching for new approaches, but only those based on the premise that the class power should remain unchanged. Which is far from easy.

The recent discussions around polycrisis indicate that large segments

86 See, e.g., Mark Landler and Stephen Castle, 'Truss, in Reversal, Drops Plan to Cut U.K. Tax Rate on High Earners', *New York Times*, 3 October 2022; Richard Partington, 'Bank Confirms Pension Funds Almost Collapsed amid Market Meltdown', *Guardian*, 6 October 2022.

of the ruling class are increasingly recognising that our turbulent times involve several interacting crises. It is likely that the ruling class and their political allies will not only aggressively *use* these crises but also deliberately combine real and pseudo-crises to *create* a heightened sense of urgency. From this it follows that only they can save us, and, on behalf of us all, they will once again reshape capitalism in their own image. It is crucial to remember that the climate crisis – the real urgency – is mainly relevant here if it can be harnessed by the ruling class to address their own hegemonic crisis.

One important reason why the ruling class and allied politicians have been able to continue down this path for so long, despite all the problems and harm it creates along the way, is because the working class is too poorly organised and the political left too weak. The analysis in this book might be reason for concern, pessimism and despair. Indeed, if the ruling class does not succeed in finding new ways of dominating workers and nature within the framework of bourgeois liberal democracies, there is always fascism. In this respect, pessimism is not irrational. But we cannot know. The only thing we can say for sure about the future is that we do not know – partly because it also depends on us.

Instead of speculating on the most likely future scenarios or describing how horrendous the future might become, I will, in the following, steer the discussion towards how we can shape it. Can we envision a socialist offensive in a world on fire?

Green New Deals with Old Grey Keynesianism?

The political left has, for decades, been craving anti-neoliberal policies. With critique of the increasing inequality stressing the need for progressive state intervention, and with the twentieth century an eternal frame of reference, it is no surprise that many again look to progressive versions of Keynesianism. The fact that China handled the 2008 crisis with the largest counter-cyclical policy the world has ever seen is viewed by many on the left (consciously or not) with some admiration. It works! That states during the 2010s threw money into the economy without causing inflation (until 2022) was certainly no detriment to Keynesianism. Let's get the money moving! Joe Biden became, in 2021, the first US president in decades to seem interested in implementing massive investments in anything other than the military or tax cuts.[87] Although Biden's investments paled in comparison to similar investments in China, and despite the familiar class character of the programmes, as a *state policy* this represented a break with neoliberalism. Was this a victory for the left, or the best way to cement ruling class power? Or maybe both?

In all the excitement over neoliberalism being on the defensive, it is easy to forget that Keynesianism is fundamentally a liberal theory. According to Altvater, it predicts that economic instability can be contained through political management. With Keynesianism, according to Lennart Schön, planning and social engineering would in many areas replace market mechanisms.[88] Keynesianism – in all its forms – seeks to cultivate the shining 'front side' of capitalism and dampen or remove the 'back side' through state policy. It is a world of private profits, liberal values and public safety nets. It is often said that Keynesianism is ultimately about saving capitalism from itself. Geoff Mann argues that John Maynard Keynes himself, in *The General Theory*, was not primarily concerned with saving capitalism or liberalism per se. Instead, the aim was rather to save the 'thin and precarious crust' of civilisation.[89] While this may be the best reading of Keynes, in practice it is not an abstract civilisation that those influenced by Keynes's ideas tend to save in crises, it is de facto capitalism. Keynesianism is, in practice, a form of crisis management where capitalist social relations are a precondition. Indeed, capitalism must be reproduced for its crisis management to function.

Variants of Green Keynesianism have in recent years emerged in the form of various *Green New Deals* (henceforth GND). These can point in a relatively progressive direction, but also help us to identify the limitations of Keynesianism. Kate Aronoff and colleagues have one of the more radical approaches to this theme. In their book *A Planet to Win: Why We Need a Green New Deal*, capitalism is ultimately incompatible with environmental sustainability, and we should replace our profit-driven energy system with democratic control over much of the

87 See Elliott, 'Biden Attempts'.

88 Elmar Altvater, *The Future of the Market: An Essay on the Regulation of Money and Nature after the Collapse of 'Actually Existing Socialism'* (London: Verso, 1993), p. 42; Lennart Schön, *En modern svensk ekonomisk historia: tillväxt och omvandling under två sekel* (Stockholm: SNS Förlag, 2000), p. 346.

89 Geoff Mann, *In the Long Run We Are All Dead: Keynesianism, Political Economy, and Revolution* (London: Verso, 2017), p. 65.

economy.[90] Naomi Klein, in *On Fire: The (Burning) Case for a Green New Deal*, launches a similar project, although perhaps slightly less radical, yet she still emphasises the importance of social movements, trade unions, scientists and local communities for the project to succeed.[91] Ann Pettifor's *The Case for the Green New Deal* is safely within a Keynesian framework and, for her, the GND is essentially a new fiscal and monetary policy.[92] Where Pettifor does not see beyond capitalism, Max Ajl's *A People's Green New Deal* advocates a full ecological and social revolution (emphasising Global North–South dimensions).[93] With Ajl, however, most connotations of the original New Deal disappear altogether. The Swedish versions of the GND are explicitly Keynesian. Katalys (with the report *Climate Keynesianism Now*) and Reformisterna (with the programme *A Green New Deal for Sweden*), both associated with a left opposition within the Swedish Social Democratic Party, are on safe intra-capitalist ground where state policy is supposed to kill two birds with one stone: restoring the welfare state and building a sustainable society. The state is the central actor, and the working class is something to be won over at elections.[94]

The GND has become so popular that even the EU has adopted programmes that play on similar strings. The goal of the European Green Deal – though without the 'New', perhaps to keep the radicals at arm's length? – is to make Europe the world's first climate-neutral region with net-zero greenhouse gas emissions by 2050.[95] For this, the EU deserves some credit, but the Green Deal has also been met with deserved criticism – for depending on absolute decoupling; for an unrealistic optimism in technology; for ignoring problems associated with the scarcity of raw materials; for using 2050 as a target; for using the concept of net emissions; and for notoriously prioritising private profit over public interest.[96] Furthermore, there is also a degrowth GND, and

90 Kate Aronoff et al., *A Planet to Win: Why We Need a Green New Deal* (London: Verso, 2019), pp. 28, 48, quote p. 74.

91 Klein, *On Fire*, pp. 259–92, esp. p. 266.

92 Pettifor, *The Case*.

93 Max Ajl, *A People's Green New Deal* (London: Pluto Press, 2021), see, e.g., pp. 94–5.

94 Kalle Sundin, *Klimatkeynesianism Nu*, Katalys, Report 83, 2020; Reformisterna, *Föreningsmotion: En grön ny giv för Sverige*, Reformisterna, 2020; Reformisterna, *Grön ny giv för Sverige*, report, Reformisterna, 2021.

95 EU, *A European Green Deal*, European Commission, 2021.

96 For discussions see, e.g., Belén Balanyá, 'How the Fossil Fuel Lobby Is Hijacking

there are radical GNDs, a feminist GND and so on.[97] That there are different GNDs should not surprise us as it is a catchy concept that can be filled with different content.

There *is* something enchanting about Green Keynesianism and the GND. Through government investment we solve economic crises, improve the situation of the broad working class, and curb global warming. If budget deficits do not particularly matter, what is not to like? But is it too good to be true? Yes and no. The neoliberal decades have almost made us forget that major social reforms are possible, even within capitalism. Many concrete reforms discussed in the framework of both the GND and Green Keynesianism are very important. That should be supported. In all the euphoria over the death of neoliberalism, we cannot turn a blind eye to some fundamental problems. Based on the analysis presented in this book, we simply cannot escape some key challenges. That the crises are rooted in capitalism carries significant implications. As a result, we must outline five challenges to Green New Deals. Let us call it a friendly critique.

First, ecological and economic crises are not solved by new investments and new technologies alone. Destruction and devaluation are at least as important. Some Keynesians would point out that they want to control investment and some GNDs want to ban the fossil fuel industry, but the main focus remains nonetheless on investments of various sorts. Although the New Deal of 1933 was politically progressive in many areas, the deep economic problems of the time were only overcome by military rearmament in the late 1930s, and the most important ingredient for 'solving' the crisis was the enormous destruction during the Second World War.[98] Even today, investments – the very essence of the

---

the European Green Deal', *Open Democracy*, 8 July 2020; Alfons Pérez, 'A Green New Deal for Whom?', *Open Democracy*, 23 April 2021.

97 Riccardo Mastini, Giorgos Kallis and Jason Hickel, 'A Green New Deal without Growth?', *Ecological Economics* 179 (2021), pp. 1–9; Carlo E. Sica, 'For a Radical Green New Deal: Energy, the Means of Production, and the Capitalist State', *Capitalism Nature Socialism* 31, no. 4 (2020), pp. 34–51; Global Feminist Frameworks for Climate Justice Town Hall, *Global Feminist Frameworks for Climate Justice Town Hall: Frameworks Reader*, September 2020. Aronoff and colleagues (in *A Planet to Win*, p. 44) also discuss what they call 'faux Green New Deal'.

98 Lars Magnusson, *Finanskrascher: från kapitalismens födelse till Lehman Brothers* (Stockholm: Natur & Kultur, 2020), p. 167; Galbraith, *The Great Crash*, p. 186. On how the 'New Deal' also reproduced racism and patriarchy, see, e.g., Pettifor, *The Case*, pp. 51–3.

GND – cannot solve our economic problems alone. Should the left mobilise on promises that it can solve economic crises through investments, when the 'best' solution would perhaps be a new world war?

Keynesians' adamant insistence that investment is the most important tool for solving both economic and ecological crises stems, I think, from two common bromides. First, that money is good: after centuries of fetishising money – where money has been the king of commodities, where everything apparently can be bought with money – is it any wonder that even socialists believe that *more* money is the solution? And second, that what is best for ordinary people is best for capitalism and vice versa. Reality is far more complex than that, in crises even more so.

Second, what do socialists do if the GND does *not* work? I do not ask this question to suggest that socialists should be intimidated from taking power or be afraid of losing. But I must ask: What if a GND fails to solve *both* crises? What if a policy based on a GND manages to reduce emissions and steer development towards a maximum of 1.5 degrees, but without creating profits or growth? When people are unemployed or have difficulty putting food on the table, it will be hard for politicians to argue for reducing emissions by another 7 per cent. Given the choice between saving the capitalist economy or the climate, to put it bluntly, there is absolutely no doubt what the ruling class will choose. The aim of the EU's deal is explicitly to make the EU a world-leading and competitive economy. Failing that, environmental policy will be put on hold. Even radical versions of GNDs might arrive at a similar destination, and even socialist politicians will be forced to make similar decisions. Socialist politicians who do not question the iron laws of capitalism will need to save the economy first, because allowing the economy to crash will have much greater consequences for people *here and now*.

Third, what does the left do if GND actually works? What would the GND look like as a regime of accumulation? If we imagine – over the next forty to sixty years – that a GND and Green Keynesianism actually manage to create economic growth *and* reduce greenhouse gas emissions: What kind of capitalism would that be? Green Keynesians point to new investments to solve both crises, but what happens *after* the big investments in railways, when the road network is electrified and new energy sources are implemented? Such investments do not create a new regime of accumulation, on a par with, say, Fordism and neoliberalism.

Discussions about Green Keynesian solutions often revolve around energy: How can we change the energy we inject into the system? But which system are we talking about? How should production, consumption, transport and infrastructure – and so on – actually be organised? It is often said that green sectors should create high growth with more well-paid jobs. It is easy to forget that the fossil fuel industry is a highly profitable sector that contributes to employment and growth, and in countries like Norway the jobs are remunerative. It's not just any old part of the economy that needs to go. We are not discussing the replacement of CFC gases in refrigerators and spray bottles; the climate crisis is of a qualitatively different nature. Those who are not interested in breaking with capitalism need to be aware of what kind of alternative regime of accumulation they are advocating for. Fewer fossil fuels in a growing capitalist economy will lead to increased pressure on nature elsewhere, on alternative energy sources and minerals and so on. This pressure will increase as the economy grows and will necessarily create new ecological crises. This also means continued dependence on exploitation and expropriation of people and nature in poorer countries.

Fourth, win–win solutions hide class. A GND *can* challenge the ruling class, it *can* seek to change power relations – but, with mainstream and social democratic variants, this is the exception rather than the rule. The refrain has been, not least in Sweden, that the state can currently afford greater debt and should borrow to finance necessary and green investments. Which may be true in itself, as many necessary investments cannot possibly be financed by higher taxes alone. When the debate is reduced to how much new money is needed and how it can be created, the main problem quickly disappears from focus: the class power that created the problems in the first place.

Fifth, and finally: strategies that do not question capitalism will reproduce it. One argument for the GND is that we can at least buy ourselves some much-needed time while opening up possibilities for deeper change.[99] This might be true, but I think (sadly) the opposite is at least as likely. Green variants of Keynesianism can be excellent ways to keep the discussion at relatively superficial levels, while reproducing the system at large. If the question is about how much it costs to stop the climate crisis, politicians can compete by promising more money. The more serious the

99 For discussion, see Boscov-Ellen, 'Infectious Optimism'.

climate effects, the more money is needed. As the climate crisis becomes worse, it may remedy the situation that we at least have active politicians who understand the seriousness of the problem and are trying to do something. Political action can prevent the ecological crisis from developing into political crises. Green Keynesianism may be the best way to *live with* the climate crisis. If Keynesianism is a way to alleviate the problems of capitalism while maintaining the core functions of the system, this may be the perfect tool to save the system and ease the pain all the way into the apocalypse.

Deals or Struggles?

One argument for Green Keynesianism, even among socialists, is that more radical ideas are unrealistic. Aronoff et al. stress that we have just over a decade to cut global carbon emissions in half, and add that we 'don't imagine ending capitalism quite that quickly'. Socialists like Rikard Warlenius and Matt Huber both have similar formulations.[100] It is interesting that the Swedish Reformisterna, who are by no means revolutionaries, nevertheless feel the need to argue that we 'don't [have] time to wait for the hypothetical collapse of capitalism'.[101] This is, of course, true, but also not the point. This obviousness – or truism – is used to legitimise a false dichotomy: either we all accept the iron laws of capitalism or we must believe in a total revolution within a few years. It took centuries to complete the transition from feudalism to capitalism; now we must *either* believe that an analogous transition out of capitalism must happen within a decade *or* we must accept the ruling order of capitalism. Faced with this dilemma, every sane person would work for green reforms rather than sit and wait for the giant collapse. But this is a straw man. It is a rhetorical figure that keep us from questioning the framework of capitalism. When revolution is defined as unrealistic, only (Green) Keynesianism remains. It is as if Rosa Luxemburg never published *Social Reform or Revolution*, as if Leon Trotsky never wrote *The Transitional Program* or André Gorz

100 Aronoff et al., *A Planet to Win*, p. 20; Rikard Hjorth Warlenius, *Klimatet, tillväxten och kapitalismen* (Stockholm: Verbal, 2022), pp. 80, 290; Matt Huber, *Climate Change as Class War: Building Socialism on a Warming Planet* (London: Verso, 2022), p. 264.

101 Reformisterna, *Grön ny giv*, p. 113, our translation.

never discussed *non-reformist reforms.*[102] Socialists do not have to choose between working for reforms or revolutions. Indeed, revolutionaries have always worked within the system while trying to transcend it.

It should not surprise us that even many radical intellectuals accept the iron laws of capitalism in the face of the climate crisis. After all, we are speaking about a *crisis.* That is, events that have traditionally put more limits on radical movements than they have opened possibilities. This must first be recognised as a problem; then we can move forward.

Beyond the straw man identified previously, the question of what is realistic is quite complicated. If we by 'most realistic' mean 'most likely', the most realistic is always not to change existing power relations. The 'most realistic' goal from this perspective is not to stop global warming at all. The better question is then: What is the most realistic way forward that could actually be implemented *and* stop the climate crisis? Or, if you like, stop ecological crises altogether? Now the total-capitalism/total-revolution dichotomy effectively conceals the most realistic alternative to confronting the climate crisis. Namely, *moving towards socialism.* Ecological crises can be met with non-capitalist reforms. Reform after reform, use-value must be prioritised over exchange-value. The economy must become more democratic and planned; land, financial systems and energy systems must be placed under democratic control and infrastructure under the logic of use-value; production must be democratically planned and workers must be empowered in workplaces. In other words: a socialist crisis policy must be developed beyond Keynesianism, reformism and capitalism.

Some key eco-socialist reforms – for example, major investment in rail transport – will sound very much like Green Keynesianism. But why is the impulse to build things that are needed or to ease the pain for the working class during a crisis always associated with Keynesianism, and not socialism? I suggest that it is mainly because there are not many developed programmes – yet – for what a truly eco-socialist crisis policy might actually look like. Investment and social reform can be as central to a socialist crisis policy as to a Keynesian one. So, what is the difference?

---

102 Rosa Luxemburg, *Social Reform or Revolution* (London: Militant Publications, 1986); Leon Trotsky, *The Death Agony of Capitalism and the Tasks of the Fourth International: The Mobilization of the Masses around Transitional Demands to Prepare for the Conquest of Power: The Transitional Program* (London: Labour Publications, 1981 [1938]); André Gorz, *Socialism and Revolution* (London: Allen Lane, 1975).

First, socialists will also fight for reforms other than those we usually associate with Keynesianism, like nationalising banks and the energy sector, or placing companies under workers' control and so on. Another difference is that Keynesianism takes capitalism for granted, which means that questions of reforms or investments must always align with questions of profit. Eco-socialism recognises that the core of the problem lies in the accumulation of capital itself, and that this must be confronted. An actually implemented socialist crisis policy needs to be pragmatic in many respects, but the starting point is this: if socialist reforms clash with the needs of the class that accumulates capital, then we adapt the economic system to the reform in question, not adapt the reform to the capitalist system.

To confront the climate crisis, we need struggles and movements, not deals. A socialist crisis policy recognises that there will be class struggle. Modern social democrats and Green Keynesians always hope to avoid this. They seem equally surprised every time reality bites back, and seek class compromises as soon as there is a clash.

Any socialist movement that genuinely succeeds in addressing ecological crises will inevitably come into conflict with private ownership. A movement that does not dare to confront the ruling class will fall flat as soon as there is a clash. Getting into the fight is a fundamental step towards solving the climate crisis. If a progressive GND is successful, it will most likely unleash dark reactionary right-wing forces allied with the ruling class. Then it is better to be prepared than to be surprised when the conflict emerges. If we – realistically! – seek to solve the climate crisis, we need to have a plan for what to do when profit and ownership stand in our way. Facing the crises of capitalism without class struggle is simply unrealistic.

Epilogue: Optimism at the Edge of a Cliff

We are on a train and the train is heading straight for the edge of a cliff. It is as if all of humanity is on the train, and it has been riding around on the tops of some Norwegian fjord for a few centuries. If this is Sognefjorden, the ice retreated about 12,000 years ago, and, for almost as long, these monumental mountains and the deep fjord have been the very definition of stability. Now, everything seems to be in flux. As the mountain shakes, the train speeds towards the cliff. On the train's public address system, the ideologues of capitalism preach to us all: look at that amazing view!

Some want the train to back up. But the iron laws of capital only force it forward. No matter how much the social democrats want to return to the 1960s or liberals to the 1990s, it is impossible to move backwards. Some dream of the state using its social engineering to conjure away the laws of capitalism; others wait for new technological revolutions to save us all. But what social engineers and technological innovators do best is speed up the train. Some pray to higher powers. Perhaps a god can redefine the law of gravity? Or miraculously incarnate a soft mattress we can all land on? Others want to isolate the different sections of the train from each other. After all, the most important thing is that brown, black and poor people do not start moving backwards to first class. Many people seem unconcerned about people falling off the cliff as long as the poor and non-whites fall first. Still others dress in tinfoil hats and try to explain that the cliff is just a conspired fairy tale. Just when we really do not have time for such foolishness, more and more people seem to fall for the idiocy. The those who are driving the train are usually silent, but every now and then, a message comes out of the loudspeakers reminding us that we are facing a great common challenge, that we will get through it together – and don't forget to enjoy the view!

Most socialists in the nineteenth and twentieth centuries would probably have invoked a different metaphor. Rather than a cliff, many would have said that the capitalist train is heading straight for socialism. The emphasis might vary – the development of productive forces, technology or social relations, whether one is more or less deterministic – but most pointed in the same direction. Forward to socialism! Many Russian revolutionaries saw their task as completing Peter the Great's work of modernising Russia. Socialism would dominate nature *better* than capitalism. Socialism was seen as a break with capitalism, which would be replaced by public ownership and a fairer distribution of resources, but it was not seen as a radical break with the kind of development that the West had experienced since the Industrial Revolution.[1]

Discourses of 'progress' still hold sway. According to the socialist eco-modernist Leigh Phillips, progress is defined as 'the steady expansion of freedom for all humanity', and it can only continue if people's power over nature continues to increase. The end of growth is, for Phillips, synonymous with 'an end to technological development, an end to science, an end to progress, an end to the open-ended search for freedom – an end to history'.[2] Keynesianism and socialist eco-modernists both carry the assumption that climate change should be dealt with without confronting the metabolic rift. Here, the problem is not the metabolic rift per se, but only how it is managed and controlled. Despite very significant differences – one imagining this within, the other beyond capitalism – both positions underestimate how constitutive the contradiction between progress and destruction is for capitalism. They are versions of the dream that one can keep the 'bright' side and eliminate the 'dark' side of the system.[3] But we

1 Enzo Traverso, *Understanding the Nazi Genocide: Marxism after Auschwitz* (London: Pluto Press, 1999), pp. 19–25. On trains and revolutions, see Enzo Traverso, *Revolution: An Intellectual History* (London: Verso, 2021), ch. 1.

2 Leigh Phillips, 'The Degrowth Delusion', opendemocracy.net, 30 August 2019.

3 Marx criticised Proudhon for dividing all economic categories into a good and a bad side, and then wanting to eliminate the bad one. Marx then asked Proudhon rhetorically: Which was the good side of slavery he wanted to keep? See Karl Marx, *The Poverty of Philosophy* (Paris: Progress Publishers, 1955 [1847]).

cannot, with the precision of a surgeon, pick out the destructive and leave the progressive.

If I were convinced that the train was heading for socialism, I too would want to join the accelerationists and speed up. But it is not. Accelerating the contradictions of capitalism does not seem to lead to socialism, but to more crises and disasters, ecological destruction and pandemics. The development of productive forces did not take us to socialism. It took us to a new geological epoch.

In a text on British rule in India, Marx compares 'progress' in the bourgeois era (capitalism) to 'that hideous, pagan idol, who would not drink the nectar but from the skulls of the slain'.[4] Even today, it is hard not to be fascinated by the construction of an alpine resort in Dubai or a fifty-seven-storey skyscraper in China built in less than three weeks. The contradiction is intertwined: it is the same progress that contributes to growth and new technology *and* degrades the health of the economy. The deep hypocrisy and inherent barbarism of bourgeois civilisation is laid bare before our eyes, Marx continued as regards British colonialism: while it assumes respectable forms at home it goes naked in the colonies. Even today, people who praise the progress of capitalism often point to specific geographical locations. Capitalism's front side presents shining apartments in sustainable seafront neighbourhoods in the Global North. The reverse side is hidden, either in neighbourhoods mired in the shadows of cities or on the other side of the planet. The contradictions of capitalism are as interconnected as capitalism is global. As the climate crisis intensifies, a socialist project must expose capitalism as naked. Anyone can point to the barbarism of a sweatshop in Bangladesh, but H&M stores selling the products – not to mention the headquarters on Mäster Samuelsgatan in Stockholm – are at least as barbaric.

The German Marxist Walter Benjamin questioned Marx's idea that revolutions were the locomotives of history.[5] Should it not be the other way around? Are not revolutions rather attempts by the passengers to pull the emergency brake? Rather than speeding up or vanishing the

4 Karl Marx, 'The Future Results of British Rule in India', *New-York Daily Tribune*, 8 August 1853.

5 See Michael Löwy, *Fire Alarm: Reading Walter Benjamin's 'On the Concept of History'* (London: Verso, 2005); Walter Benjamin, *The Arcades Project* (Cambridge, MA: Harvard University Press, 1999).

precipice by sorcery, perhaps we should try to stop the train? The socialist revolution would be about stopping the headlong rush to disaster. Benjamin wrote that capitalism would not die a natural death, and there is no reason to think climate change will kill capitalism any day soon either. The capitalist train will neither stop by itself nor accelerate into socialism. It will run straight off the precipice if we do not pull the emergency brake.

Drawing on Benjamin's thinking, Enzo Traverso points out that many who continued to see capitalism as a productive force on the road to socialism had difficulty understanding the genocide of Europe's Jews. For them, this was a kind of return to pre-modern and primitive forms of society. In fact, the Holocaust was a modern capitalist 'barbarism': high-tech, industrial, efficiently administered and thoroughly organised, based on rail transport meticulously managed, and on the latest theories and academic research. After Auschwitz, Zygmunt Bauman argued, we must question our culture because it was imagined and realised within the framework of our civilisation, not at its borders but in one of its most developed parts. Nicos Poulantzas showed how fascism was never a reactionary undeveloped form of capitalism. It represented a development of the capitalist forces of production, with industrial development, technological innovations and an increase in the productivity of labour: 'all the while promoting the *expanded reproduction* of the conditions of capitalist production, that is, reinforcing class exploitation and domination'. Fascism, with Poulantzas, was not a '*backward turn*, but rather a *forward rush*'. In light of Auschwitz, Hiroshima and Kolyma, Traverso writes, the choice is no longer between socialism and the decline of humanity, but rather between a socialism perceived as a *new civilization* and the *extinction of* humanity.[6]

The crises of capitalism are no more exceptions from 'modernity', 'progress' and 'civilization' than colonialism, the Holocaust or sweatshops. The latest crisis is always the freshest and most modern thing capitalism can offer. Crises are not exceptions to progress or breaks from modernity. The crises *are* capitalism, progress and modernity.

6 Traverso, *Understanding the Nazi Genocide*, p. 25; Nicos Poulantzas, *Fascism and Dictatorship: The Third International and the Problem of Fascism* (London: Verso, 2018), pp. 98–100; see also Ugo Palheta, 'Fascism, Fascisation, Antifascism', *Historical Materialism*, 7 January 2021, historicalmaterialism.org.

The watchword in the wake of the First World War for Rosa Luxemburg and the Spartacists was 'Socialism or Barbarism'. Luxemburg was building on Engels who argued that bourgeois society faced a difficult dilemma: 'either an advance to socialism or a reversion to barbarism'.[7] The slogan 'Socialism or Barbarism' has been interpreted in different ways, but, from the history of crises, we see that barbarism is certainly a component *of* capitalism. With the climate crisis, we can take this one step further: barbarism is what we get if capitalism does *not* go under. Engels wrote that we might fall back into barbarism. This is not true. We advance towards it. When we cross the two-degree warming threshold, we will be riding around on new high-speed trains with the latest version of air-conditioning and using new technology to read about extreme wildfires in the US or floods in Pakistan.

Ecological crises are highly modern. They have historically followed the latest innovations, with new technologies, more profitable energy systems and a more efficient division of labour. Always with careful, rational calculations. Highly trained staff and brilliant scientists and engineers have used the latest and greatest knowledge to dig up more and more oil and coal. Many people associate new technology with green and environmentally friendly energy. They forget that the fossil fuel industry is still a highly technological industry, and that big-tech companies often have massive climate impacts. A global infrastructure for transporting oil, coal, energy and physical goods was made possible by clever spatial planning and geographical analysis. The climate crisis was created not by idiots looking back, but by highly educated people seizing tomorrow.

The climate crisis – like the Holocaust – is surrounded by an icy rationality. Today, highly educated people sit and calculate the exact cost to the world economy of 1.5 or 2.4 degrees of warming. It is like putting a price on poor people's heads. Does it 'pay' to carry out the sixth mass extinction of species on Earth? Is it 'worth it' to let the polar ice caps melt? Barbarism, Traverso writes, is inextricably linked to the civilisation we live in.[8] The climate crisis is the civilisation *and* barbarism of our time.

7 Rosa Luxemburg, *The Crisis in the German Social-Democracy (The 'Junius' Pamphlet)* (New York: Socialist Publication Society, 1919), p. 18; Norman Geras, *The Legacy of Rosa Luxemburg* (London: Verso, 2015), ch. 1.

8 Traverso, *Understanding the Nazi Genocide*, p. 106.

Challenging the idea of progress is no small task when our modern history is so deeply rooted in the very belief in progressive advancement. It also means getting rid of the belief that capitalism has always given socialists a tailwind. Socialism is not the logical continuation of capitalism; socialism is a *break* with capitalism. This is not some magical metaphysical break, but simply an organisation of the economy according to people's needs (use-value) rather than profit (exchange-value). It is a socialism that is qualitatively – not only quantitatively – different from capitalism. Crises are not opportunities that bring us closer to our goal; they are problems we must fight. *We must stop them.*

To take control of the train, we must confront the class that drives and owns the train. They are actually on the train too, but in a first-class cabin at the back. Socialism means pulling the emergency brake, taking control and stopping the train. Then we get off the train and start walking. We turn ninety degrees from the precipice and walk together to the left. We walk quickly, but in an organised fashion. That is how we leave this hellish mountain.

Ten Thoughts on Socialist Crisis Policy

But 'pull the emergency brake'? What does it actually mean? Pulling the emergency brake cannot be a mere abstraction or radical rhetoric. It cannot be a romanticisation of pre-capitalist conditions, nor an illusion that we can start from scratch. Socialism must be a break with capitalism, but what does it really mean when the world is on fire? As a philosophical statement, it is interesting and relevant. But, in the face of actual crisis, we need to be way more pragmatic. Refusing to invest in improved technology in the midst of a crisis because one is against modernity would be as absurd as refusing to think beyond 'growth' because one thinks progress equals freedom. It is as flawed to vehemently oppose *all* high-speed trains merely because they are symbols of modern industrialisation as it is to wholeheartedly support *all* such trains under the assumption that only by outcompeting air travel can we address the climate crisis. When shifting focus from crisis critique to crisis policy, we must become more practical and pragmatic. There will always be various forms of continuity, whether we like it or not. What we break with will inevitably influence the future. Socialism may be a *break* with capitalism, but it is still a break with *capitalism.*

One major challenge when developing a socialist *crisis policy* is to address that crises are crises. Crisis exists, by definition, on several levels of abstraction; so must socialist approaches to the crisis. We must tackle both the immediate suffering and the underlying causes. A socialist approach must be relevant to those affected, broad enough to gain power and deep enough to confront the causes of the problems. As outlined in the introduction, we need a general socialist *crisis critique* to understand the nature and history of capitalist crisis, to grasp the terrain we need to walk, to show how capitalism produces crises and how crises produce capitalism. But a movement that merely exists on this level is politically useless. We therefore need *crisis policies* that can guide us when crises hit; these must be prepared in advance so we already have an idea of what to do when a crisis emerges. Without these, we will over and over again be caught in a Keynesian fishing net where we implement policies to ease social pain while reproducing many of the processes that created the crisis in the first place. But we can never know exactly what the next crisis looks like – thus neither exactly what to do – before it has actually emerged. So, we need a socialist *crisis management*: short-term and hands-on actions and politics to handle the phases of crisis as shocks and panics, and alleviate suffering from day one in the crisis. Developing these three levels – *critique*, *policy*, *management* – and not least making them fit together, must be a joint effort between socialist activists, politicians and intellectuals.

Uniting such different levels during a crisis is certainly not an easy task. Still, Keynesians and neoliberals have actually been quite good at this. When the crisis hits, there is a knee-jerk response among Keynesians: increase state spending and investments! This can be implemented in a variety of ways, but the reflex is there, and it does relate to an overall understanding of how the political economy works. For neoliberals, the reflex has been – at least until recently – to privatise, deregulate, lower taxes and bail out banks. This reflex could also take on a number of other forms, but was always related to an overall understanding of the economy. Socialists are too often missing this reflex, which ties an overall socialist analysis together with the *here and now* in the midst of a crisis. (A major difference here, which makes the task harder for socialists, is that Keynesianism and neoliberalism are top-down projects intended to reorganise capitalism, while socialism is a movement from below seeking to move beyond the existing system.)

Our main focus has been *crisis critique*, but we will close the book with ten thoughts on a socialist crisis policy. I must admit that I seldom think books benefit very much from ending with points on what to do, but some things in life are more important than writing good books. My thoughts on policy come with several caveats. They are quite general. The analysis is (still) limited to economic and ecological crises. Analyses that point beyond capitalism must also include further reflections on what socialism is; that requires more knowledge and more theories than I offer here, not least on social reproduction and people's everyday lives. The main emphasis in this book has been on the capitalist class; other books could have started from the perspective of the (heterogeneous) working class. I hardly discuss questions of political subjects. My emphasis is on policies and political programmes; how to actually mobilise for this is at least as important, but not the main focus here. These ten thoughts must therefore be read in parallel with other thinking and action going on elsewhere.

1) Socialist Crisis Policy Is a Bridge between Present and Future: For a Transitional Programme

Socialism as a crisis policy cannot be based on what was possible yesterday, or what will be impossible even tomorrow. Strategies and programmes that merely describe idealistic circumstances in an unspecified future are often called maximum programmes, and strategies for tomorrow that include such small and circumscribed requirements that they can simply be taken along for the ride down the precipice are called minimum programmes. The communalist Debbie Bookchin argues explicitly that today we need both minimum *and* maximum programmes. She says nothing about the relationship between the two. That is, nothing about a path between local political demands and the dream of a completely different world.[9] Today, the mixture of short-term problem-solving and utopian idealism is exactly what is *not* needed. Rather, we

9 See, e.g., Debbie Bookchin, 'The Future We Deserve', in Barcelona en Comu (ed.), *Fearless Cities: A Guide to the Global Municipalist Movement* (Oxford: New Internationalist, 2019), pp. 12–16.

must find links and transitions between the crises as events and underlying structures. When crises come with their shocks and panics, socialists must have a programme to turn to. So-called transitional programmes were conceived in their time as a critique of minimum and maximum programmes and are more relevant today than ever. Eco-socialist transitional programmes are needed to plan, build and organise a new hegemony, to connect everyday demands and spontaneous anger with longer-term socialist goals.[10] It is only when the crisis hits that we know exactly what it looks like this time, but when the world is on fire it is difficult to plan for longer-term strategies. Transitional programmes are bridges between crisis critique and crisis management, and must be prepared *before* the crisis strikes.

2) Class: The Richest Should Pay for the Crisis

In every crisis, the poor and workers – in tandem with their experience of racism and imperialism, patriarchy and sexism – are expected to pay through their suffering. When the crisis hits, we *know* there will be an attack on working-class wages, jobs, houses, rights and so on. A socialist crisis policy has a very different class character. Not only must the working class defend their position, a socialist crisis policy must be class struggle in the opposite direction. Those who have the most must pay the most when the crisis comes. Wealth accumulated in private hands in good times should be used in economic crises to save jobs; wealth accumulated by destroying the planet must be used to mitigate the damage produced. The next step is to confront ownership, so often hidden behind aesthetics and culture or drowned in some stratification analysis.

Socialism confronts ownership. Companies struggling or going bankrupt in crises could be taken over by workers, cooperatives, municipalities or the state. If a direct takeover is not feasible, states and municipalities must intervene and facilitate the establishment of

10 On transitional programmes, see Leon Trotsky, *The Death Agony of Capitalism and the Tasks of the Fourth International: The Mobilization of the Masses around Transitional Demands to Prepare for the Conquest of Power: The Transitional Program* (London: Labour Publications, 1981 [1938]); Rosa Luxemburg, 'Our Program and the Political Situation', in Peter Hudis and Kevin B. Anderson (eds), *The Rosa Luxemburg Reader* (London: Monthly Review Press, 2004), pp. 357–72.

similar companies with a different class character. In a broader context, this also means expropriating land and essential industry, nationalising banks and financial systems, and so on. How to actually do this will depend on what the concrete crisis looks like, but it is hard to imagine this happening on a significant scale if it has not been discussed and planned before the crisis actually hits.

The crisis comes with class hatred. Every so often, some writer, not seldom with personal experience of class mobility from low-paid menial jobs to low-paid cultural jobs, writes about their rage against the rich. Upon which the infected debate returns: Is class hatred acceptable? At the same time, every crisis is flooded with class hatred in the opposite direction. Class hatred that finds concrete expressions in labour markets, housing markets and politics, that is incorporated and concealed in racism and legitimised in the media. Now and again, class hatred upwards is articulated in a novel. Class hatred downwards accompanies each crisis when it is taken for granted that the lives of workers and the poor are the ones that must be shattered. We have become so accustomed to the class hatred of the bourgeoisie that it feels natural that the poor must die in the crises of capitalism. The question today is not whether class hatred from the bottom up is legitimate, but whether there is anything else that can save us from total barbarism.

3) For Socialist Creative Destruction: Destroy What Destroys the Planet

Climate policies focus largely on investments; but to reduce emissions, it is more important that we *destroy* than that we build. And destruction is political. Politicians can choose internal devaluation (austerity, privatisations, wage cuts) or external devaluation (currency devaluation), let sectors go bankrupt or geographical areas fall into disrepair, or simply start wars. The Rehn–Meidner model of post-war Sweden, for example, was a politically driven process of destruction, where less productive sectors were to be regulated away through wage policies. A socialist crisis policy must raise the question: What can eco-socialist creative destruction look like?[11] We must start by destroying one particular sector of

11 Ståle Holgersen and Rikard Warlenius, 'Destroy What Destroys the Planet: Steering Creative Destruction in the Dual Crisis', *Capital and Class* 40, no. 3 (2016), pp. 511–32.

capitalism: namely the fossil fuel industry. This goal comes with quite a challenge. Where Lenin aimed to break the *weakest* links of imperialism and capitalism, we must confront the system's arguably strongest link. A very cautious first step is to stop subsidising fossil capital, while the necessary second step is to start banning it. This means destruction of a class fraction, and devaluation of the fossil capital infrastructure (which could trigger, deepen or perhaps solve economic crisis).

We can never invest our way out of major crises, but we must invest. Destruction is perhaps most important, but new investments are needed. This means new infrastructure, new train lines and new ports, new housing and new factories in new places, perhaps even new cities, new food production and distribution systems, and new energy systems. And so on. Socialists cannot think we are on the offensive every time the state intervenes in crises; virtually all parties advocate this. But socialist crisis policies must include monetary, fiscal and investment policies, not because these alone can solve the crises, or should save the system, but because – if properly targeted – they can ease the pain of the working class. They can also – if directed precisely – point the way forward in a new direction.

4) Socialism Must Confront Growth, but We Cannot Start by Stopping It

The degrowth movement deserves all possible credit for putting economic growth on the agenda.[12] The naïveté still prevalent among many socialists, namely the idea that ecological crises can be solved simply by decoupling and acceleration, must be rejected. Nor is the steady state economy possible within capitalism. Economic growth is surely a problem. But, in economic crises, we see the consequences of slow economic growth under capitalism, and it is not pretty: unemployment, poverty and misery. Most degrowthers do not, of course, advocate simply a capitalism without growth, but rather a completely different society. Still, I cannot see how one can mobilise the working class during economic crises with slogans that scream *less growth*.

12 For discussions, see Giorgos Kallis, Christian Kerschner and Joan Martinez-Alier, 'The Economics of Degrowth', *Ecological Economics* 84 (2012), pp. 172–80; Matthias Schmelzer, Andrea Vetter and Aaron Vansintjan, *The Future Is Degrowth: A Guide to a World beyond Capitalism* (New York: Verso, 2022). On post-growth as alternative to degrowth, see Nancy Fraser and Rahel Jaeggi, *Capitalism: A Conversation in Critical Theory* (Cambridge: Polity, 2018), p. 184.

We do not have to choose between accelerationism and degrowth. Socialist class struggle in the 2020s must acknowledge that we cannot – contra to so-called socialist eco-modernism – have infinite increase in economic activity on a limited planet. Certainly not centuries of increase in biophysical throughput. But neither can we – contra to degrowth – mobilise the broad working class or any broad movement by making 'less growth' the focal point of our project. Fortunately for us, we do not have to choose between those who see 'growth' as synonym for freedom and those who see 'growth' as the main enemy. Rather than the dichotomy of being 'for' or 'against' growth, a socialist crisis policy needs critical discussions on which sectors, places and industries should have more economic activity and which must be shut down. Establishing new 'green jobs' or sustainable infrastructure will indeed result in increased economic growth (as measured by an increase in GDP) in the short term, which obviously cannot be an argument against such policies. These are complex questions, but an eco-socialist movement seeking to mobilise beyond niche intellectual circles must provide concrete, place-specific answers to these kinds of questions.

The goal of socialism is not more or less growth. The Franco-Brazilian Marxist Michael Löwy stressed in 2015 the importance of looking beyond anti-growth (the degrowth movement) and pro-growth (eco-modernists and accelerationists).[13] He was absolutely right. The concept of 'more or less' economic growth relies on a quantitative approach to the phenomenon, but socialism is a *qualitative* breach – it is a different kind of development. If *de*growthers think their project is about qualitative, and not quantitative, measures, they should consider changing slogans. A socialist and democratically planned economy cannot be measured by the yardstick of capitalism. The fight against an economy based on eternal exponential economic growth cannot come out of a direct confrontation with 'growth', but must come from a collective and long-term socialist project – planning and building for use-value over exchange-value. We must confront it by confronting the system that creates it.

13 Michael Löwy, *Ecosocialism: A Radical Alternative to Capitalist Catastrophe* (Chicago: Haymarket, 2015), p. 32.

5) Within and beyond the System: Eco-socialist Policy as Non-reformist Reforms

Marxists have always worked both within and beyond capitalism. *The Communist Manifesto* is more than anything a call for revolution, but, as soon as the class struggle is concretised, reforms like progressive income tax and free public education pop up even here.[14]

There is a difference between reform and reformism. The Austrian French philosopher André Gorz famously distinguished between reforms that preserve the system (reformism) and reforms that challenge and point away from the system (non-reformist reforms).[15] Climate compensation and aviation tax are examples of the former, where money raised can be used for something progressive while the system itself legitimises continued flying. Demands for bread, land and peace may be limited demands in many situations, but in the case of the Russian Revolution it was precisely such demands that became revolutionary. Whether reforms actually become reformist or revolutionary is difficult to say in advance, as this also depends on the political situation and social struggles. Non-reformist reforms prioritise use-value over exchange-value, and are determined by what is humanly necessary, not by what benefits the economy or is politically and economically feasible in the short term.

Reformism has played a major role in history; that time is not now. Social democratic reformism became dominant in a context of high growth (which perhaps only a world war can create), the absence of deep economic crises, what climate scientists call 'the great acceleration', and an environment that threatened the bourgeoisie with revolutionary socialism. *These are not our times.* Post-war reformists were in no hurry, and they accelerated a climate crisis along the way. Now we are very much in a hurry, and we need to start *repairing* ecological disasters along the way. Social democrats in Scandinavia have moved away from a position of strength in the class struggle and missed (or ignored) the fact that the winds of class struggle have shifted and class compromise is

14 Karl Marx and Friedrich Engels, *The Communist Manifesto* (London: Pluto Press, 2008); China Miéville, *A Spectre, Haunting: On the Communist Manifesto* (London: Head of Zeus, 2022), pp. 57–9, 88–95.

15 André Gorz, *Socialism and Revolution* (London: Allen Lane, 1975), ch. 3; André Gorz, *Strategy for Labor: A Radical Proposal* (Boston: Beacon Press, 1968), pp. 9–28.

over, but cling to the notion – or desire – to be a party of power. Those same social democrats have ended up with reformism without reforms, and a lost chance to build socialism. On the road to the Anthropocene, socialist parties cannot make the same mistakes again.

Is 1.5 degrees a non-reformist reform? How much patience will today's youth have with their rulers? What will today's youth be doing in ten years if our rulers have continued to fail to solve the climate crisis? What excuses will politicians trot out as temperatures hurtle towards 2 degrees? It is not inconceivable that more and more people will demand that the 1.5- or 2-degree target should be met, *regardless* of the economic consequences. That we reorganise the economy rather than play Russian roulette with the climate. Can a political project that actually stops global warming at 1.5 or 2 degrees become some sort of a non-reformist reform? A key concern for Gorz when he wrote about reforms in the 1960s, after decades of relatively few crises, was that non-reformist reforms could potentially *generate* crises. Today, we can safely say that we do not need to create instability or crises. Those are already here. While Gorz wanted to provoke crises, the climate crisis takes on such proportions that even the greatest madman would not want it.

6) A Socialist Crisis Policy Must Be Conducted in and against the State

We cannot escape the state. This is complicated, as capitalist states in general seek to maintain capitalism, states can pursue the absolute worst possible policies, and state ownership is never an aim in itself. Still, we need some pragmatism here. There are, of course, major political and geographical differences among states, but, in general, it is very hard to imagine dramatic changes in the political economy within a decade or two without any access to state power. Crisis management is primarily located at state level, and the political left must fight over this. States can be used for both destruction and investment: they can ease the pain and nationalise key parts of the economy. Before a crisis hits, we must discuss what a state should demand in return when it rescues companies or sectors during the crisis.

A socialist crisis policy will not be a bed of roses. It is not forbidden to dance or to dream at the edge of the cliff, but the most important thing is that we walk away from it, together, in an organised and determined manner. Socialism in our time is not about abolishing all power relations overnight. That may be bad news for utopians, but our revolution will not be a dance straight into paradise. It will be a tortuous path to save the lives of the world's poor and protect human life on Earth. Socialism in our time is a rescue mission. Anarchism may have its time in human history, but it is not now. In a world of accelerating ecological crises where a ruling class is only growing more powerful, there is no escaping the fact that power and oppression must be used to move the world in a socialist direction. Socialism will be very much about confiscation and prohibition. Ownership must be regulated. Fossil capital must be banned. Luxury consumption too. No individual can have Tellus as their playground anymore, much less outer space. Socialism in our time will not only be about liberating ourselves, it will be at least as much about dismantling the privileges of the ruling class. Lenin makes a distinction between confiscation and socialisation. To do the first, according to Lenin, one needs 'only determination', but socialisation is something else; here other qualities are needed.[16] We must confiscate to stop the climate crisis; we must socialise to heal the metabolic rift.

7) Eco-socialism Turns Capitalist Geography Upside Down

For many accelerationists, the solution to the world's problems is to make the economy more global; for many deep ecologists, the basic rule is that everything must become very local. (In addition, we can mention Stalinists who still believe that everything should be managed by the nation state.) Our analysis should not start with a geographical scale. Rather, we need to start with two questions: How do we best organise the economy for people and the environment, and what should be produced where, for whom and how? The answers will give us the geography. The question of where to locate hospitals, housing, factories and energy production, for example, must begin with principles of use-value and the good of people and nature. A research programme may put forward a perfect model for the Nordic countries or for Europe, or for the world. It will of course – fortunately? – never be fully realised because of all sorts of local politics and local conditions. This is, nonetheless, a democratic and ecological *starting point* for a new geography. As there is

16 V. I. Lenin, '"Left-Wing" Childishness', in *Lenin's Collected Works*, vol. 27 (Moscow: Progress Publishers, 1972), pp. 323–34.

no new magic technology that will allow everyone to travel or consume like a global minority does today, I think it is fair to argue that eco-socialism will be *more* regional and local. Not as a principle, but (most likely) as a conclusion. New – or old – energy sources like solar, wind and water are, interestingly, more difficult to transport. Now, we have yet to see what the global capitalist class can do with these energy sources, but, if it becomes easier for municipalities, cities, neighbourhoods, regions and countries to produce energy themselves, there is at least the potential for a more locally based economy.

Socialism produces concrete places based on use-value, not abstract spaces based on exchange-value. The ability to transport coal and oil anywhere, and thus to locate production where it was most profitable, has been a central element in production of modern capitalist landscapes. We do not know what the geography of eco-socialism will look like, but it cannot be based on imports of cheap goods or labour, or other injustices that characterise global capitalism. Socialist crisis policies cannot escape questions of international solidarity, climate debt, uneven ecological and economic development and issues of poverty, inequality and development needs in the Global South. Discussions about the *what*, *how* and *where* of production open doors to greater democratisation of society. Democracy is about sharing power, which is precisely what happens when economic decisions are moved out of closed rooms with unelected owners.

We must invert the geography of capitalism. An eco-socialist geography is concrete, not abstract; it is based on use-value, not exchange-value; it is democratic, not capitalist. When the ruling class produces crises in its own image, we get more or less the exact opposite of what we really need. We get global crises from a non-democratic global economy organised by exchange-value – always underpinned by nationalist politics and racism. What we need is a democratic and more locally organised political economy backed up by global solidarity. We need to invert the geography of capitalism: what is global must become more local; what is national must become global.

8) Socialism against Racism and Fascism

Any socialist approach to crises must work from the assumption that the next crisis will be accompanied by racism. Exactly how racialised groups, immigrants or religious minorities will be blamed and attacked we do not know – but it will happen. The extent to which crises are mediated and lived through ethnicity, culture or 'race' is determined by political and social struggles. This has a massive impact on the development of the crises, and it is a struggle we must not lose. Anti-racism and anti-fascism are necessary components of the class struggle, just as class struggle must be equally necessary for anti-racism and anti-fascism.

The climate crisis may foster fascism. When hegemonic crisis remains unsolved and the economy seems to be in a jam, there is always an opportunity for the ruling class to look to the far right. The fact that states are (pro)active and dominant in the political economy will be welcomed and used by the far right. For them, the nation is the political answer to everything, from unemployment and insecurity to crime and environmental policies. At the same time as we confront alliances between the fossil fuel industry and the climate-denialist far right, we must also be aware of green nationalism or so-called eco-fascism. Eco-fascism may use aesthetics and propaganda to mobilise around the symbolism of unspoiled nature and picturesque landscapes, but as a political regime, fascism is always a way of organising modern capitalism and thus destroying nature.

Anti-racism and anti-fascism are defensive and offensive – even in crisis. Confronting extreme nationalism, anti-feminism and racism is partly a defensive struggle. But defensive victories are only temporary; they last only until the next round begins. Therefore, we also need struggles that confront the crises of capitalism. When Marx said that people do not make their history in circumstances of their own choosing, he could not have known that our circumstances would include a different geological epoch. The circumstances that surround us are anything but optimal. The crises are systematically allowing some to die an untimely death so that others in a distinct but tightly interconnected political geography can accumulate to an imagined infinity.

9) Life against Capital

In a 1985 text, Ernest Mandel replaced the slogan 'Socialism or Barbarism' with 'Socialism or Death'. He was outraged by the fact that 16 million children were dying every year from starvation and curable diseases. The price of capitalism's survival, according to Mandel, was several world wars against children between 1945 and 1985 alone.[17] The killing did not stop in 1985. Since then, capitalism has evolved further, grown larger, modernised even more, and produced new technologies – the internet, mass air travel, mobile phones, self-driving cars and robots. All this as corporations increasingly emphasise *corporate social responsibility*. Today, billionaires engage in space tourism while children die from poverty.

Eco-socialist crisis policies put health and life at their centre. Slogans like 'Capital against Life' rightly took centre stage during the Covid-19 pandemic, and care workers and social reproduction finally became central to the class struggle, largely thanks to sound feminist analyses. In context of *crisis critique*, we must continue to expose the causes of the pandemics and the metabolic rift. In context of *crisis policy*, we must plan how to prevent as many future pandemics as possible. This means supporting small-scale agriculture, monitoring ecological niches, stopping global warming, bringing health and social care back under democratic control, and, crucially, democratising the production of medicines and vaccines.

Social reproduction is social revolution. The working class will, one hopes, one day become the gravediggers of capitalism. But there is no reason to assume that 'productive' workers have greater revolutionary potential than other members of this class. The history of class struggle is full of social reproduction issues regarding health, housing, food and so on. An eco-socialist crisis policy must be feminist and prioritise life over capital accumulation. The coronavirus pandemic showed that consultants, financial speculators and influencers do not engage in necessary occupations. Underpaid health workers do. When promoting non-reformist reforms with the hope of mobilising the working class, we must base them on analyses of what daily life can best look like – the use-value of everyday life. This will obviously include an analysis of how to organise work, but also families and neighbourhoods and caring.

17 Michael Löwy, 'Ernest Mandel's Revolutionary Humanism', in Gilbert Achcar (ed.), *The Legacy of Ernest Mandel* (London: Verso, 1999), pp. 24–37.

10) Falsify This Book

If some of the core arguments in this book are correct – that capitalism survives because, and not despite, its crises; that crises are opportunities for the capitalist class, and dangers for workers and the political left – there are indeed reasons for pessimism. Considering that there will be more crises tomorrow, it might be difficult to escape feelings of hopelessness. To be honest, I would certainly have preferred to come to different conclusions. I would like to say that crises are opportunities for workers and that socialists must embrace crises and hope for more and deeper ones in the near future. 'World history would indeed be easy to make', Marx wrote in a letter to Kugelmann during the Paris Commune of 1871, 'if the struggle were to be taken up only on condition of infallibly favourable chances.'[18] The crises of capitalism do not work to our advantage. And, if crisis and capitalism constantly reproduce each other, where could hope and optimism even possibly come from?

Another world is possible – despite *the crisis.* When you finish this book, I propose that you put it aside and look at people and the world in a broader perspective than what we have discussed here. While I have been working on this book, there have been huge protests all over the world: from Lebanon, Iraq, Syria, Spain, France, Chile, Ecuador and Hong Kong, to Black Lives Matter and a climate movement that re-emerged as soon as corona restrictions were lifted. With the pessimism of the intellect, we see that crises reproduce capitalism; with the optimism of the will, we see a world of solidarity and love among people, and anger and hate against exploitation, injustices and extractivism. We know that a vast majority would benefit from a change in political economy, and from history we know that angry people can mobilise and unite around political projects. There is indeed fertile ground even for some intellectual optimism.

Exceptions do happen. When the ruling class under capitalism rules through crises, it becomes our task to make an exception to this rule, and exceptional things do happen. C. L. R. James aptly wrote about the unexpected nature of revolutions: 'On 22 January [Lenin] cautioned

18 Karl Marx, 'Marx to Ludwig Kugelmann in Hanover', in *Marx and Engels Correspondence* (New York: International Publishers, 1968 [1871]), available at marxists.org

that he might not live to see the revolution. Yet in March, it was there. I wonder if you get the significance of that. That is what happened to Lenin. You never can tell. Marx phrases it like this, "The revolution comes like a thief in the night."'[19] The everyday life of class struggle comes through daily reforms, but struggle also comes with sudden upsurges and events that are discontinuous. Openings for exceptions – to create socialism – will come, certainly not every day, but they will come. There is much we cannot know about revolutions, but one thing we know is that revolutions actually do happen. Often, we cannot know about historical possibilities until they have actually materialised. If not aware and prepared, we will not even know when the socialist potentials are here. If we remain committed to reformism and intra-capitalist solutions in every crisis, such opportunities might pass without anyone even noticing. In a world on fire, we must start winning against the crises of capitalism. This will take us into new territory, where we do not know what the political landscape will look like. On a positive note: at least we know that new territories do occur in history.

Like most authors, I want to be read and liked. I want people to think I wrote clearly and had something interesting to say. That I 'was right'. With this particular book, there is something else I want much more. We need to *make* my general conclusions *untrue*. I want the main arguments – that crises reproduce capitalism and mainly benefit the capitalist class – to become politically irrelevant. We must change reality so that librarians will be able to move this book from the political section to the historical one. Let us outdate my book. Let us falsify my thesis.

19 C. L. R. James, *Walter Rodney and the Question of Power* (London: Race Today Publications, 1982).

Index

Note: Page numbers in **bold** indicate **tables**.